U0100078

零失敗
秘方系列

簡易快煮好味湯餸

Quick and easy tasty meals

編者話

Preface

快煮，滿足急促節奏；慢吃，領略美食真味！

你我都是忙碌的職場都市人，打理完公務，拖着倦極的身軀回家，肚子餓餓怎麼辦？

對自己的身體好一點，美食是穩定情緒的好物。

走進廚房，拋卻日間煩瑣，可嘗試炮製書內的冷盤、蒸餸、小炒、烤物、鍋品及速湯，30 分鐘內，溫暖的美食填滿空洞的肚，感覺滿足多了！書內指導多項快速烹調的小技巧，例如食物先解凍、醃味，或取用預先準備的醬料，只需炒炒煮煮，美食瞬間完成。

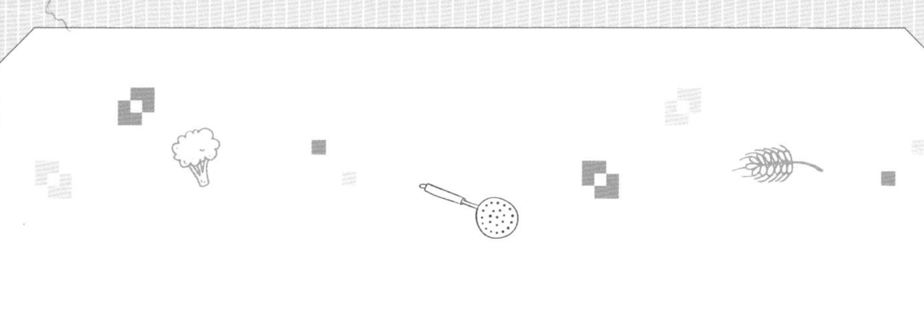

Cook fast, without sacrificing taste or nutritional value.
Eat slow, to appreciate the true flavours.

All urbanites are busy workers. After a tough day in the office, you drag your exhausted body home. What should you do to feed yourself properly?

You deserve a good meal. Good food always makes you feel happy.

Go straight into the kitchen and forget about your abrasive boss or annoying co-workers. This cookbook includes appetizers, steamed dishes, stir-fries, grills, casseroles and quick soups. That means you're never more than 30 minutes away from home-made comfort food that nourishes the body and soul. Each recipe comes with tips on fast cooking, such as thawing, marinating or making sauces ahead of time. You too can make yourself a tasty dinner in just a few simple steps.

目錄

Contents

教你快煮小竅門 / 6
Tips on Fast Cooking

冷盤 · 小吃
Appetizers and Snacks

麻辣青瓜 / 8
Cold Cucumber Appetizer Dressed in Sichuan Pepper Chilli Oil

白烚豬頸肉 / 12
Poached Pork Cheek

毛豆煮烤麩 / 14
Braised Deef-fried Gluten with Edamame Beans

酸辣四季豆沙律 / 17
Sour and Spicy Snap Bean Salad

小炒 · 蒸餸
Stir-fries and Steamed Dishes

洋葱牛肉炒番茄 / 20
Stir-fried Beef with Onion and Tomatoes

金銀檸檬蒸黑鯧魚 / 22
Steamed Black Pomfret with Salted and Fresh Lemon

乾煸鴕鳥肉 / 24
Sautéed Ostrich Meat

黃金芙蓉魚米 / 27
Stir-fried Fish Dices with Corn Kernels and Egg Whites

麵醬蒸龍躉腩 / 30
Steamed Giant Grouper Belly in Soybean Paste

蝦醬蒸豬頸肉 / 32
Steamed Pork Cheek with Shrimp Paste

黑椒百合牛柳條 / 35
Stir-fried Beef Tenderloin and Lily Bulb in Black Pepper Sauce

鴛鴦棗蒸文昌雞 / 38
Steamed Wenchang Chicken with Black and Red Dates

蝦醬炒魷魚筒 / 40
Stir-fried Baby Squids in Shrimp Paste

麵豉啫啫唐生菜煲 / 42
Sizzling Chinese Lettuce Casserole with Soybean Paste

煎焗
Fried Food and Grills

椒鹽雞中翼 / 44
Fried Mid-joint Chicken Wings with Peppered Salt

蕉葉烤魚 / 46
Baked Sole Fillet with Banana Leaf

焦糖香葱燒排骨 / 49
Caramel Roasted Pork Spareribs

煎三文魚柳 / 52
Fried Salmon Fillets

豆漿芝士焗西蘭花 / 54
Baked Broccoli with Soybean Milk and Cheese

照燒汁鯖魚 / 56
Fried Mackerel in Teriyaki Sauce

燜煮 · 鍋物
Stewed Dishes and Casserole

九層塔煮雞球 / 58
Three-cup Chicken with Thai Basil

紅酒牛柳芥蘭鍋 / 60
Beef Tenderloin Casserole with Kale and Red Wine

簡易沙薑汁浸雞髀 / 62
Chicken Legs in Spice Ginger Sauce

枝竹酸菜魚腩煲 / 65
Grass Carp with Tofu Stick and Pickled Vegetable in Clay Pot

日式咖喱蘋果雞 / 68
Japanese Apple Chicken Curry

豆漿鮮淮山雞肉鍋 / 70
Yam and Chicken Nabemono with Soybean Milk

鮮蝦麻婆豆腐 / 72
Stir-fried Tofu with Shrimps

三葱焗雞 / 74
Stewed Chicken with Assorted Onions

滾湯
Quick-boiled Soups

杞子枸杞菜魚尾湯 / 76
Fish Tail Soup with Qi Zi and Wolfberry Vine

節瓜鹹蛋湯 / 78
Chinese Marrow Soup with Salted Egg

香芹胡椒豆腐蜆湯 / 80
Clam Soup with Tofu and Chinese Celery

紫菜蘿蔔肉碎味噌湯 / 82
Miso Soup with Seaweed, White Radish and Pork

木棉魚大豆芽番茄湯 / 84
Big-eye Fish Soup with Soybean Sprout and Tomato

芫茜皮蛋魚片湯 / 86
Grass Carp Soup with Coriander and Century Egg

蝦仁雪耳羹 / 89
Shrimp and White Fungus Thick Soup

芥菜排骨湯 / 92
Sparerib Soup with Mustard Greens

韓式海鮮湯 / 94
Korean Seafood Soup

菠菜豬膶湯 / 96
Pork Liver Soup with Spinach

苦瓜鹹菜海魚湯 / 98
Fish Soup with Bitter Melon and Salted Mustard Green

紅蘿蔔竹蔗馬蹄湯 / 100
Water Chestnut Soup with Carrot and Sugarcane

洋葱番茄薯仔牛肉湯 / 102
Beef Soup with Onion, Tomato and Potato

鹹菜桂花魚湯 / 104
Chinese Perch Soup with Salted Mustard Greens

冬瓜海鮮湯 / 106
Winter Melon and Seafood Soup

草菇肉絲豆腐羹 / 108
Tofu Thick Soup with Straw Mushroom and Pork

淮山紅菜頭葉素菜湯 / 110
Vegetable Soup with Beetroot Leave and Yam

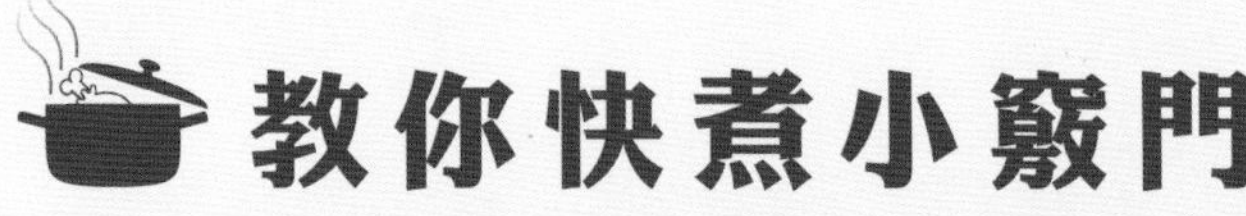

教你快煮小竅門
Tips on Fast Cooking

要有效地短時間煮一餐，只要懂得適當地運用煮食技巧及方法，下班後煲湯煮餸不再是麻煩事，反而令你大有滿足感！

準備篇

- 需要解凍的肉類或海鮮，宜早一晚放於雪櫃下層自然解凍，翌日可即時煮吃。
- 預計需長時間醃製的食材，可前一天醃味，免卻當天醃製時間不足。
- 選擇周末或周日一次性地選購食材，除可節省成本外，以免下班後緊迫的時間。
- 煮好醬料放於雪櫃，如番茄醬、肉醬、南瓜汁等，可隨時取用配搭各款材料。

煮食篇

- 預早煮好冷盤小食，晚餐當天取出即可食用，成為晚餐的其中一道美食，節省烹調時間。
- 用多款材料煮成的鍋物，最適合時間緊湊的都市人，一鍋有菜有肉的熱辣美食，滿足晚餐的營養要求。
- 妥善運用多種廚具幫忙，同一時間蒸、燜、焗、炒，以免浪費時間。

滾湯篇

- 由於煲湯時間少，建議材料切薄或切成小塊，容易入味、煮熟。
- 魚類是滾湯常用的材料，鮮甜美味，而且所需時間不多，海魚及淡水魚皆可。
- 可加入瓜菜、肉類、魚、海鮮等材料，一併喝湯、吃料及伴飯吃，毋須烹調其他餸菜，成為簡單的一餐。

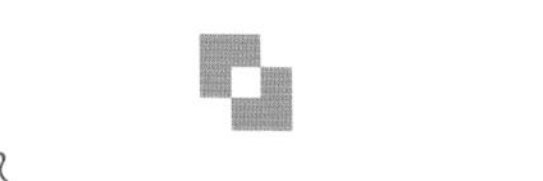

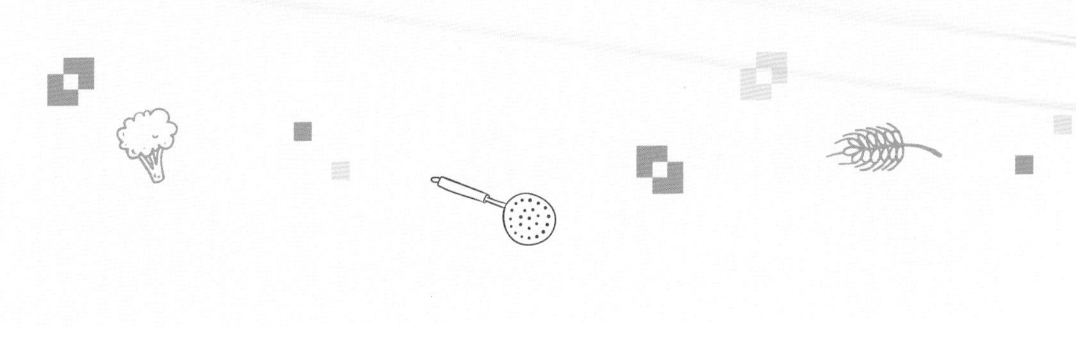

To create a meal within a short period of time, you need to cook and prepare the food wisely. It's not troublesome at all to make soup after work if you know the tricks. You may even derive much satisfaction from it.

Preparations

- All meat or seafood that need to be thawed should be moved from the freezer to the refrigerator the night before for slow and safe defrosting. You can use it the following night right away.
- Ingredients that take time to marinate should be marinated one night ahead, so that they have enough time to pick up the seasoning.
- Shop for grocery in bulk once a week on Saturday or Sunday. You can save some money that way and you don't need to waste time on grocery shopping after work on weeknights.
- Make a range of sauces in advance and store them in the fridge. Sauces such as tomato sauce, meat sauce or pumpkin sauce work on many different ingredients. They instantly add flavours and depth to any food you make.

Cooking

- Make cold appetizers ahead and refrigerate them. Pick one and serve as one course on weeknights. That saves you much time.
- Make a casserole with various ingredients. This is the perfect one-pot quick fix for any urbanite pressed for time. You'd get veggies, meat and starch in one pot, fulfilling your nutritional requirements.
- Make good use of cooking tools. Use one stove for steaming or stewing and the other for stir-frying while another dish is being baked in the oven. That's how you save time.

Quick-boiled soups

- As the ingredients are not cooked for too long in the soup, you should dice them or slice them finely, so that the flavours are infused and the ingredients are cooked more quickly.
- Fish is a great ingredient for making quick soups. It takes just a short while to cook and it tastes rich and flavourful. You may use marine fish or freshwater fish for soups.
- Add meat, veggies, fish or seafood to the soup. Serve the soup with rice and eat the solid ingredients also. That makes a meal in itself and you don't need to make any other dishes.

麻辣青瓜

Cold Cucumber Appetizer Dressed in Sichuan Pepper Chilli Oil

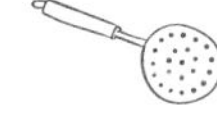

材料

溫室小青瓜半斤
雲耳 1/3 兩
蒜茸 2 茶匙

醃料

麻香辣椒油 2 茶匙（參考 p.10）
麻油 2 茶匙
鹽、糖各半茶匙

做法

1. 雲耳用水浸軟，去硬蒂，洗淨，飛水，過冷河。
2. 小青瓜洗淨，切去頭尾兩端，用刀拍裂，再切成塊。
3. 將青瓜塊、雲耳、蒜茸及醃料放於容器內拌勻，冷藏後即可食用。

Ingredients

300 g hothouse baby cucumbers
13 g cloud ear fungus
2 tsp finely chopped garlic

Marinade

2 tsp Sichuan pepper chilli oil (refer to p.10)
2 tsp sesame oil
1/2 tsp salt
1/2 tsp sugar

Method

1. Soak the cloud ear in water until soft. Cut off the tough roots. Rinse well. Blanch in boiling water. Rinse with cold water.
2. Rinse the cucumbers. Cut off both ends and crush them with the flat side of a knife. Then cut into pieces.
3. Put the cucumber, cloud ear fungus, garlic and marinade into a container. Mix well. Refrigerate briefly. Serve.

1

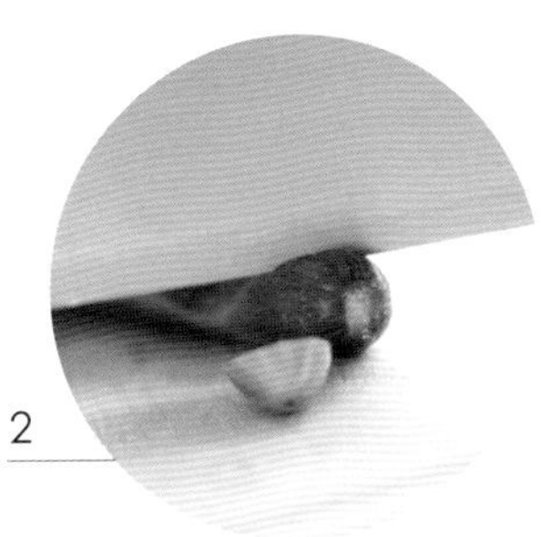
2

3

4

麻香辣椒油

Sichuan Pepper Chilli Oil

材料

川椒粒 4 湯匙
指天椒 4 兩
豆豉 1 湯匙
蝦米 2 湯匙
乾葱茸 2 湯匙
蒜茸 1 湯匙
粟米油 1.5 杯

調味料

鹽 1 茶匙
糖 1 茶匙
生抽 1 湯匙

做法

1. 蝦米洗淨，切碎；豆豉用水沖洗，切碎；指天椒洗淨，去蒂、切碎。
2. 燒熱油 1.5 杯，下川椒粒用小火炸至散發香味，隔去大部份川椒粒，下蝦米、指天椒及乾葱茸，用小火炒至帶香味，加入豆豉及蒜茸炒勻，最後下調味料煮 5 分鐘（整個炒煮過程約 15 至 20 分鐘），待涼，入瓶儲存。

Ingredients

4 tbsp Sichuan peppercorns
150 g bird's eye chillies
1 tbsp fermented black beans
2 tbsp dried shrimps
2 tbsp finely chopped shallot
1 tbsp finely chopped garlic
1.5 cups corn oil

Seasoning

1 tsp salt
1 tsp sugar
1 tbsp light soy sauce

Method

1. Rinse the dried shrimps and chop them finely. Rinse the fermented black beans with water and then chop them finely. Rinse the bird's eye chillies. Cut off the stems and chop finely.
2. Heat a wok and pour in 1.5 cups of oil. Fry the Sichuan peppercorns over low heat until fragrant. Set aside most of the Sichuan peppercorns with a strainer ladle. Add dried shrimps, bird's eye chillies and shallot. Stir fry over low heat until fragrant. Add fermented black beans and garlic. Stir well. Add seasoning at last and cook for 5 minutes. (The whole cooking step takes about 15 to 20 minutes.) Leave it to cool. Transfer into sterilized bottle.

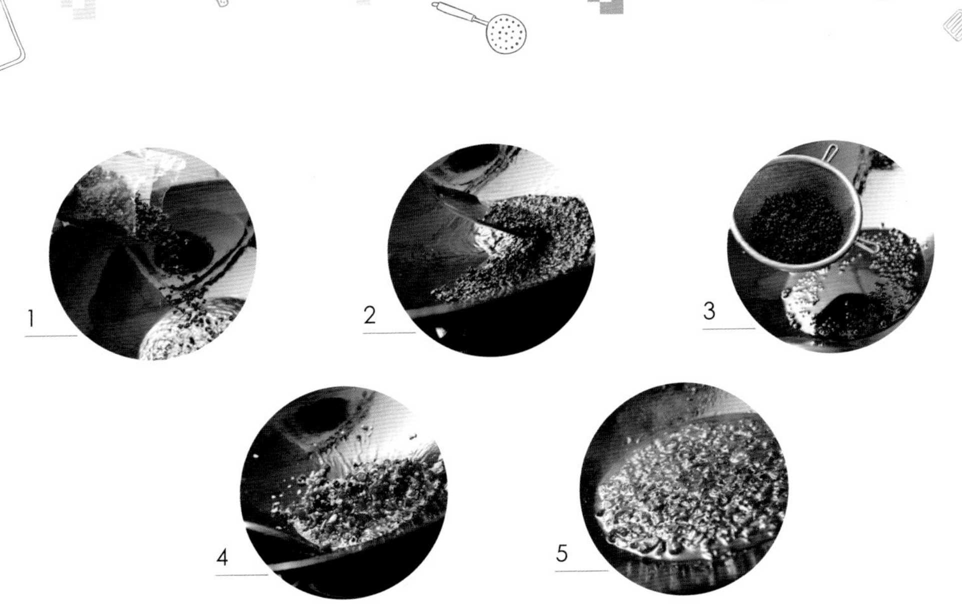

零失敗技巧
Successful cooking skills

為何雲耳先飛水？

這是涼拌菜，建議雲耳飛水煮透，以去除菇菌的霉味。

Why do you blanch the cloud ear fungus first?

This is a cold appetizer. Thus, it's advisable to blanch the cloud ear fungus to cook it through, so as to remove its mouldy taste.

為何青瓜用刀拍裂，而非切塊？

用刀拍裂的青瓜，咬入口質感佳，而且帶鬆脆的口感。

Why do you crush the cucumber, but not cutting it?

Crushing the cucumber with the flat side of a knife makes it crack along its natural grain. The cucumber tends to have better mouthfeel with a lovely crunch.

炒煮麻香辣椒油，有何注意之處？

炒川椒粒時，必須注意火候，用小火炒透，別冒出大煙及炒焦，時間太久令油帶苦澀味。此汁放於雪櫃可儲存半年，炒煮或蘸汁皆可。

Is there anything that needs my attention when I make the Sichuan pepper chilli oil?

When you stir-fry the Sichuan peppercorns, pay attention to the heat you use. Do it over low heat without burning them or making them smoke. Otherwise, the chilli oil will end up tasting bitter. This dressing lasts in the fridge for 6 months. You can use it as a dip or a condiment for other dishes.

白焓豬頸肉

Poached Pork Cheek

30 分鐘
烹調時間

材料

新鮮豬頸肉 2 塊（約 12 兩）
潮州鹹菜心 3 兩
薑 3 片
八角 2 粒
玫瑰露酒 1 湯匙

蘸汁（調勻）

紅椒碎半茶匙
蒜茸 1 茶匙
鹽 1/3 茶匙
糖半茶匙
米醋 4 湯匙

做法

1. 豬頸肉洗淨，燒滾清水半鍋，放入豬頸肉、薑片及八角，煮滾，轉小火煲半小時，至豬肉全熟，傾入玫瑰露酒煮至大滾，盛起豬頸肉，待涼。
2. 鹹菜心洗淨，切絲。豬頸肉斜切成薄片，上碟，伴鹹菜絲及蘸汁享用。

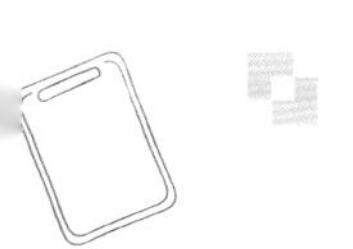
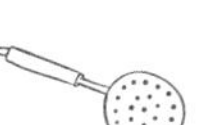

Ingredients

2 pieces fresh pork cheek (about 450 g)
113 g Chaozhou pickled mustard green
3 slices ginger
2 star anise
1 tbsp Chinese rose wine

Dipping sauce (mixed well)

1/2 tsp chopped red chilli
1 tsp finely chopped garlic
1/3 tsp salt
1/2 tsp sugar
4 tbsp rice vinegar

Method

1. Rinse the pork cheek. Bring 1/2 pot of water to the boil. Put in the pork cheek, ginger and star anise. Bring to the boil. Turn to low heat and cook for 1/2 hour until it is fully done. Pour in the Chinese rose wine and bring to a rolling boil. Dish up. Allow it to cool down.
2. Rinse and shred the pickled mustard green. Slice the pork cheek diagonally. Remove on the plate. Serve with the pickled mustard green and dipping sauce.

零失敗技巧
Successful cooking skills

如何得悉豬頸肉全熟？
豬頸肉轉成啡色，而且刺入沒有血水滲出即全熟。
How to know the pork cheek is cooked through?
It is done when it turns brown and no blood is oozing out when pierced.

用白焓的方法，火候必須用小火嗎？
用小火慢煮，能夠徹底保留肉香，而且肉質嫩滑。
Does poaching require low heat?
Gently cooking the meat over low heat can fully retain its flavour and give a smooth texture.

已加有八角及薑片的香味，為何還要加入玫瑰露酒？
減低豬頸肉的肉腥味，而且增添酒香，吃起來酒香撲鼻！
The star anise and ginger are already fragrant. Why add Chinese rose wine?
It is to reduce the raw smell of the pork cheek and enhance the flavour with the wine fragrance. It smells wonderful!

毛豆煮烤麩
Braised Fried Gluten Ball with Edamame Beans

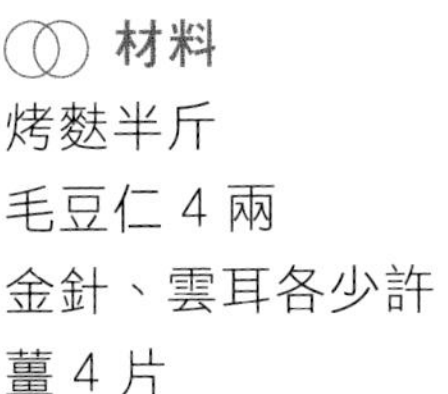

材料

烤麩半斤
毛豆仁 4 兩
金針、雲耳各少許
薑 4 片

調味料

蠔油 2 湯匙
生抽 1 湯匙
麻油、糖各半湯匙
胡椒粉少許
水 1.5 杯

做法

1. 金針、雲耳用水浸軟，剪去硬蒂，飛水。
2. 毛豆仁洗淨，飛水。
3. 烤麩用手撕成小塊，啤水片刻，用手壓乾水分，飛水，過冷河，再壓乾水分。
4. 燒熱鑊，下油 1/3 杯，加入烤麩炒至微黃，盛起，用滾水略沖，壓乾水分。
5. 原鑊留下油 1 湯匙，炒香薑片，加入金針、雲耳、毛豆仁及烤麩炒勻，下調味料煮滾，轉慢火煮 20 分鐘，至汁液濃稠即成。

Ingredients

300 g fried gluten balls
150 g edamame beans
dried lily flowers
cloud ear fungus
4 slices ginger

Seasoning

2 tbsp oyster sauce
1 tbsp light soy sauce
1/2 tbsp sesame oil
1/2 tbsp sugar
ground white pepper
1.5 cups water

Method

1. Soak the dried lily flowers and cloud ear fungus in water until soft. Cut away the hard ends. Scald.
2. Rinse edamame beans and scald.
3. Tear the fried gluten balls into small pieces by hand. Rinse under the tap running water for a while. Squeeze dry. Scald and cool in cold water. Squeeze dry.
4. Heat a wok. Put in 1/3 cup of oil. Stir-fry the fried gluten balls until lightly brown. Dish up. Lightly rinse with boiling water. Squeeze dry.
5. Leave 1 tbsp of oil in the wok. Stir-fry the ginger until fragrant. Add the dried lily flowers, cloud ear fungus, edamame beans and gluten. Stir-fry and mix well. Add the seasoning and bring to the boil. Turn to low heat and cook for 20 minutes until the cooking sauce reduces. Serve.

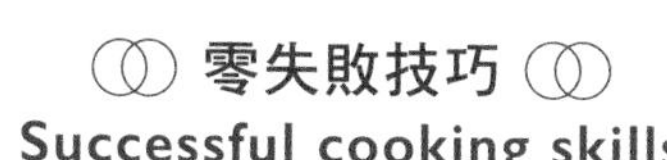

零失敗技巧
Successful cooking skills

烤麩用手撕開有何好處？
保留烤麩的質感，以免刀切破壞口感。
Why tear fried gluten balls into pieces by hand?
It keeps the texture of gluten better than cutting with a knife does.

烤麩啤水有何作用？
有效去掉烤麩發酵時的酸味。
What is the purpose of rinsing fried gluten balls under running water?
It can remove the sour taste of gluten during fermentation.

處理烤麩有何要訣？
必須徹底壓乾水分，令烤麩乾爽，以免炒煮後糊成一團。
What is the tips for cooking fried gluten balls?
Squeeze the gluten dry to avoid pasting together during cooking.

如何先預備，放工後可馬上烹調？
可預早一天將烤麩啤水、過冷河及壓乾，冷藏待翌日炒煮即可。
Is there any preparation that I can do ahead of time, so that I can make this dish right away after work?
You may prepare the fried gluten balls one day ahead. Just rinse them under a slowly running tap. Rinse in cold water and squeeze them dry. Keep them in a fridge and you can use them right away the next day.

酸辣四季豆沙律
Sour and Spicy Snap Bean Salad

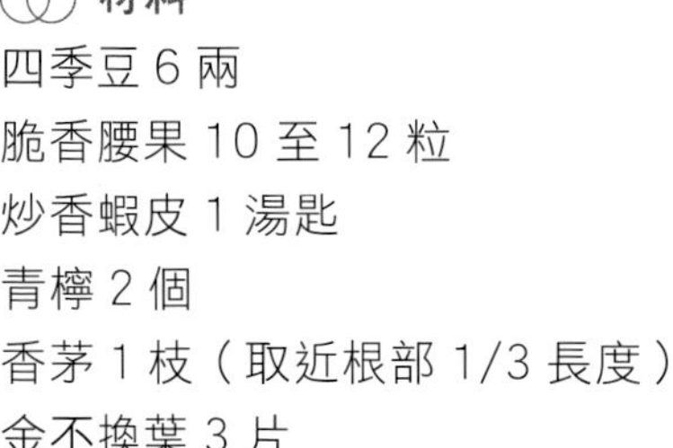

材料

四季豆 6 兩
脆香腰果 10 至 12 粒
炒香蝦皮 1 湯匙
青檸 2 個
香茅 1 枝（取近根部 1/3 長度）
金不換葉 3 片
指天椒 2 隻
蒜肉 1 粒

做法

1. 青檸壓出汁，備用。
2. 香茅、金不換葉、指天椒、蒜肉全部洗淨，切碎，放入石盅內，用木槌搗成茸，盛起，加入青檸汁及調味料拌成泰式酸辣汁。
3. 四季豆撕去老筋，洗淨，切段，放入滾水灼 3 分鐘，盛起，放入冰水過冷河，瀝乾水分。
4. 四季豆放於碟上，澆上酸辣汁，最後加入腰果及蝦皮食用。

Ingredients

225 g snap beans
10 to 12 crisp cashew nuts
1 tbsp dried tiny shrimp (stir-fried)
2 limes
1 stalk lemongrass (only 1/3 in length near the root taken)
3 Thai basil leaves
2 bird's eye chillies
1 clove skinned garlic

Method

1. Squeeze out the lime juice. Set aside.
2. Rinse the lemongrass, Thai basil leaves, bird's eye chillies and garlic. Chop all up, put into a stone mortar and crush into puree with a wooden pestle. Dish up. Add the lime juice and seasoning, mix together to become the Thai sour and spicy sauce.
3. Tear the strings off the snap beans, rinse and cut into sections. Put into boiling water and blanch for 3 minutes. Put into iced water to cool and drain.
4. Put the snap beans on a plate and pour the sour and spicy sauce on top. Finally add the cashew nuts and tiny shrimp. Serve.

◎ 零失敗技巧 ◎
Successful cooking skills

為何將香茅等材料先切碎、後搗成茸？

這樣香茅及金不換葉等可徹底散出香味，而且搗碎後混和較易掌握，毋須太費勁。

Why the lemongrass and other spices are chopped up before crushed into puree?

It is to fully release the aroma of lemongrass, Thai basil leaves and other spices. It is also easier to mix them together.

泰式酸辣汁可預早調勻嗎？

為了省時方便，可預先搗碎並混和，冷藏後可隨時使用，方便上班一族。

Can I make the sour and spicy sauce ahead of time?

Yes, you can. To save time, you can crush the ingredients and mix them in advance. Refrigerate it and you can use it to dress your favourite salad right away. This works especially for busy workers.

洋葱牛肉炒番茄

Stir-fried Beef with Onion and Tomatoes

20 分鐘
烹調時間

材料

番茄 2 個（約半斤）
洋葱半個
新鮮牛肉 4 兩
蒜肉 2 粒（拍鬆）

醃料

胡椒粉少許
生抽半湯匙
水 2 湯匙
粟粉 1 茶匙
油 1 湯匙（後下）

調味料

鹽 3/4 茶匙

做法

1. 牛肉洗淨，切片，下醃料拌勻醃 2 小時，備用。
2. 洋葱撕去外衣，洗淨，切絲；番茄洗淨，去蒂，切片。
3. 燒熱鑊下油 2 湯匙，下洋葱炒香盛起。
4. 原鑊下油 1 湯匙，下蒜肉爆香，加入牛肉炒至轉色，放入番茄、洋葱及調味料，炒片刻即成。

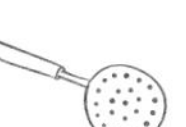

Ingredients

2 tomatoes (about 300 g)
1/2 onion
150 g fresh beef
2 cloves skinned garlic (crushed)

Marinade

ground white pepper
1/2 tbsp light soy sauce
2 tbsp water
1 tsp cornflour
1 tbsp oil (added at last)

Seasoning

3/4 tsp salt

Method

1. Rinse and slice the beef. Mix with the marinade and rest for 2 hours. Set aside.
2. Peel the onion. Rinse and cut into shreds. Rinse and remove the stalk from the tomatoes. Cut into slices.
3. Heat up a wok. Add 2 tbsp of oil. Stir-fry the onion until fragrant. Dish up.
4. Add 1 tbsp of oil in the same wok. Stir-fry the garlic until scented. Put in the beef and stir-fry until it changes colour. Add the tomatoes, onion and seasoning. Stir-fry for a moment. Serve.

零失敗技巧
Successful cooking skills

購買哪部份牛肉，肉質最軟脍？
牛柳的肉質軟脍，容易咀嚼，也適合長者食用。

Which part of the beef is most tender?
Tenderloin is soft and gentle and easily to be chewed. It is suitable for elders.

如何才能吃到番茄的香甜味道？
最後才放入番茄炒煮，以免番茄的香味經久煮而流失。

What can I do to retain the aromatic sweetness of the tomatoes?
Put in the tomatoes at last. The aromas of tomatoes tend to disappear after prolonged cooking.

金銀檸檬蒸黑鯧魚

Steamed Black Pomfret with Salted and Fresh Lemon

10 分鐘
烹調時間

材料

黑鯧魚 2 厚片（約 12 兩）
鹹檸檬半個
新鮮檸檬 3 片
熟油 1 湯匙
葱絲少許

醃料

胡椒粉少許
粟粉 1 茶匙

做法

1. 鹹檸檬去核，剁碎，備用。
2. 黑鯧魚洗淨，抹乾水分，下醃料塗勻，鋪上鹹檸檬茸及鮮檸檬片，隔水大火蒸 6 分鐘至全熟，澆上熟油，伴上葱絲即成。

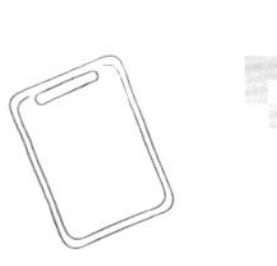

Ingredients

2 thick slices of black pomfret (about 450 g)
1/2 salted lemon
3 slices fresh lemon
1 tbsp cooked oil
shredded spring onion

Marinade

ground white pepper
1 tsp cornflour

Method

1. Seed the salted lemon. Finely chop it. Set aside.
2. Rinse the black pomfret. Wipe dry. Add marinade and spread well. Arrange chopped salted lemon and fresh lemon slices on top. Steam over high heat for 6 minutes until fully done. Drizzle with smoking hot oil. Garnish with shredded spring onion. Serve.

零失敗技巧
Successful cooking skills

配上金銀檸檬，食味有何特別？
鹹檸檬的鹹香；鮮檸檬的鮮香，兩者完全滲入魚肉，是伴飯的最佳菜式。

What's special about using both salted and fresh lemon?

Salted lemon tastes savoury and aromatic, while fresh lemon tastes sour and refreshing. The intense flavours of both are picked up by the fish. It is a great dish to go with rice.

醃料為何加入粟粉？
粟粉可封鎖魚塊的肉汁，保持汁液及鮮味。

Why do you marinate the fish with cornflour?

It seals in the juice of the fish, keeping it succulent and flavourful.

用黑鯧魚塊來蒸，有何好處？
避免太多碎骨，進食時啖啖肉，適合長者及小朋友食用。

Why do you use thickly sliced black pomfret for this recipe?

Black pomfret is fleshy and does not have small bones, so that this dish is great for all ages – including kids and seniors.

乾煸鴕鳥肉
Sautéed Ostrich Meat

材料

急凍鴕鳥肉 200 克
洋葱半個（切幼絲）
紅蘿蔔 1/4 個（切幼絲）
中芹 2 棵（去葉，切度）
蒜茸 2 茶匙
薑絲 1 湯匙
辣豆瓣醬 1.5 茶匙
磨豉醬 1 茶匙
紹酒 1 茶匙

醃料

鹽及糖各半茶匙
生抽、紹酒及粟粉各 1 茶匙
麻油及胡椒粉各少許
水 1 湯匙

調味料

鹽 1/4 茶匙
糖 1 茶匙
生抽 1 茶匙
粟粉 1 茶匙
水 2 湯匙
麻油及胡椒粉各少許

做法

1. 鴕鳥肉放於雪櫃下層自然解凍，洗淨，抹乾水分，切成幼條，下醃料拌勻。
2. 燒熱油 1 湯匙，下洋葱絲、蒜茸及薑絲略炒，加入鴕鳥肉略煎兩面，推散輕炒。
3. 加入辣豆瓣醬及磨豉醬炒勻，灒紹酒，下紅蘿蔔絲及中芹絲拌勻。
4. 最後加入調味料拌炒片刻，上碟享用。

Ingredients

200 g frozen ostrich meat
1/2 onion (finely shredded)
1/4 carrot (finely shredded)
2 stalks Chinese celery (leaves removed; sectioned)
2 tsp finely chopped garlic
1 tbsp shredded ginger
1.5 tsp chilli bean sauce
1 tsp ground bean sauce
1 tsp Shaoxing wine

Marinade

1/2 tsp salt
1/2 tsp sugar
1 tsp light soy sauce
1 tsp Shaoxing wine
1 tsp cornflour
sesame oil
ground white pepper
1 tbsp water

Seasoning

1/4 tsp salt
1 tsp sugar
1 tsp light soy sauce
1 tsp cornflour
2 tbsp water
sesame oil
ground white pepper

Method

1. Defrost the ostrich meat in the lower chamber of the refrigerator. Rinse and wipe dry. Cut into thin strips. Mix with the marinade.
2. Heat up 1 tbsp of oil. Slightly stir-fry the onion, garlic and ginger. Add the ostrich meat and fry both sides lightly. Scatter and stir-fry gently.
3. Add the chilli bean sauce and ground bean sauce. Stir-fry evenly. Sprinkle with Shaoxing wine. Put in the carrot and Chinese celery. Mix well.
4. Finally add the seasoning and stir-fry for a moment. Serve.

零失敗技巧
Successful cooking skills

切鴕鳥肉有何技巧？

鴕鳥肉緊記順橫紋切條，並用洋葱、蒜茸及薑絲起鑊，能享受美味細緻的鴕鳥肉。

Is there any trick in slicing ostrich meat?

When you slice ostrich meat, make sure you cut it across the grain into strips. Stir-fry onion, garlic and shredded ginger until fragrant before you put it in the wok. That's the trick to tender and delicious ostrich meat.

如何將鴕鳥肉炒得嫩滑？

鴕鳥肉的肉質比牛肉更容易處理，只要略煎兩面，推散略炒，鴕鳥肉嫩滑好吃。另外，加入少許水分醃味，可增加鴕鳥肉的滑嫩口感。

How to make stir-fried ostrich meat soft and gentle?

Ostrich meat is easier than beef in treatment. Slightly fry both sides; scatter and then slightly stir-fry. The ostrich meat will be tender and flavourful. Also, adding a little water to marinate the meat will enhance the smooth taste.

這道菜的要點是甚麼？

這道菜的烹調時間毋須太久，緊記快速兜炒，而且先下洋葱爆香，令鴕鳥肉帶洋葱的甜香味。

What is the cooking main point of this dish?

This dish does not need to be cooked for too long. Remember to stir-fry quickly and stir-fry the onion first until it is fragrant. It gives the ostrich meat a sweet onion flavour.

黃金芙蓉魚米

Stir-fried Fish Dices with Corn Kernels and Egg Whites

材料

魚柳 4 兩
新鮮粟米粒 3 湯匙
蛋白 5 個
芫茜 1 棵（切碎）

醃料

鹽及雞粉各 1/4 茶匙
胡椒粉少許
生粉半茶匙
油 1 茶匙

調味料

清雞湯 1/3 杯
鹽 1/4 茶匙
糖 1/8 茶匙
麻油及胡椒粉各少許
生粉半湯匙

做法

1. 魚柳洗淨，抹乾水分，切粒，加入醃料待 5 分鐘。
2. 粟米粒用滾水煮約 2 分鐘至熟透，盛起待涼。
3. 熱鑊下油，下魚米炒至半熟，待涼。
4. 蛋白拂勻，加入調味料、魚米及芫茜拌勻。
5. 熱鑊下油，注入蛋白混合物及粟米粒炒至剛凝固，上碟，趁熱享用。

Ingredients

150 g fish fillet
3 tbsp fresh corn kernels
5 egg whites
1 stalk coriander (finely chopped)

Marinade

1/4 tsp salt
1/4 tsp chicken bouillon powder
ground white pepper
1/2 tsp caltrop starch
1 tsp oil

Seasoning

1/3 cup chicken broth
1/4 tsp salt
1/8 tsp sugar
sesame oil
ground white pepper
1/2 tbsp caltrop starch

Method

1. Rinse fish fillet and wipe dry. Dice and marinate for 5 minutes.
2. Cook corn kernels in boiling water for about 2 minutes until done. Let them cool.
3. Add oil into a hot wok. Stir-fry diced fish until medium cooked. Let it cool.
4. Whisk egg whites. Mix in the seasoning, fish and coriander.
5. Add oil into a hot wok. Pour in the egg white mixture and corn kernels. Stir-fry until just set. Serve hot.

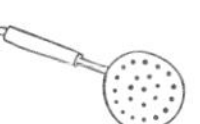

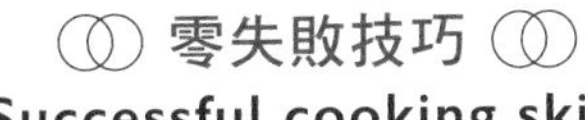

零失敗技巧
Successful cooking skills

應選擇哪款魚柳切成魚米？

魚柳可選用斑肉、青衣肉或龍脷柳等。

Which kind of fish fillet should be chosen to make fish dices?

Grouper, green wrasse or sole fillet can be used.

用甚麼火候炒蛋白魚米？

炒魚米時用中慢火，令蛋白魚米呈雪白色澤，又不焦燶。

What level of heat should be used to stir-fry the fish dices?

Stir-fry fish dices over medium-low heat can give the egg white and fish dices snow white gloss and also not charred.

如何有效地剝出完整之粟米粒？

先剝掉一直行之粟米粒，其餘的就能輕易及完整地剝出。

How to tear out whole corn kernels effectively?

Tear away a line of corn kernels then the remaining can be easily torn out in whole.

可用罐裝粟米粒代替嗎？

當然可以，但清甜味當然不及新鮮粟米粒。緊記粟米粒必須徹底吸乾水分，才與蛋白同炒，否則菜式會滲出水分，影響味道。

Can it be replaced with canned corn kernels?

Of course but their taste is not as fresh and sweet as that of fresh corn kernels. Remember to wipe dry the corn kernels thoroughly before stir-frying with egg whites or the dish becomes watery.

麵醬蒸龍躉腩

Steamed Giant Grouper Belly in Soybean Paste

材料

生劏龍躉腩 12 兩
麵豉醬 1 湯匙
薑 6 片
芫茜 4 棵
熟油 1 湯匙

醃料

胡椒粉少許
粟粉 1 茶匙

調味料

生抽半湯匙
糖 1/4 茶匙

做法

1. 芫茜去掉鬚頭，洗淨，切段。
2. 龍躉腩洗淨，斬件，下醃料拌勻醃 10 分鐘。
3. 麵豉醬與調味料拌勻，加入龍躉腩拌勻，上碟，鋪上薑片，隔水大火蒸 10 分鐘，放上芫茜，澆上熟油即成。

Ingredients

450 g slaughtered giant grouper belly
1 tbsp fermented soybean paste
6 slices ginger
4 stalks coriander
1 tbsp cooked oil

Marinade

ground white pepper
1 tsp cornflour

Seasoning

1/2 tbsp light soy sauce
1/4 tsp sugar

Method

1. Remove the root of the coriander. Rinse and cut into sections.
2. Rinse the giant grouper belly. Chop into pieces. Mix with the marinade and rest for 10 minutes.
3. Mix the soybean paste and seasoning. Add the giant grouper belly and mix well. Put on a plate. Line the ginger on top and steam over high heat for 10 minutes. Top with the coriander. Sprinkle with the cooked oil and serve.

零失敗技巧
Successful cooking skills

用麵豉醬及生抽調味，吃到魚香嗎？

當然會！因為麵豉醬及生抽的份量不多，而且龍躉魚肉的味道非常鮮濃！

Can we taste the real flavour of the fish seasoned with fermented soybean paste and light soy sauce?

Of course! Only a small amount of the seasoning is used and the fish is already rich in flavour by itself.

如何去掉魚腥味？

徹底洗掉龍躉腩的血水，以免腥味掩蓋魚鮮味！

How do you remove the fishy taste of the grouper belly?

Make sure you rinse off all the blood on the fish. Otherwise, it covers up the authentic flavours of the grouper belly.

蝦醬蒸豬頸肉
Steamed Pork Cheek with Shrimp Paste

10 分鐘

烹調時間

材料

豬頸肉 6 兩
布包豆腐 1 塊
陳皮 1 角
紅辣椒 1 隻（切圈）

醃料

蝦醬 1 茶匙
生抽 1 茶匙
紹酒 2 茶匙
糖半茶匙
粟粉 1 茶匙

做法

1. 陳皮用水浸軟，刮淨內瓤，切絲。
2. 豬頸肉洗淨，斜切成薄片，下醃料拌勻待 1 小時。
3. 布包豆腐洗淨，切片，排於蒸碟內，鋪上豬頸肉，灑入陳皮絲及紅椒圈，隔水大火蒸 10 分鐘，趁熱食用。

Ingredients

225 g pork cheek
1 cube cloth-wrapped tofu
1 quarter dried tangerine peel
1 red chilli (cut into rings)

Marinade

1 tsp shrimp paste
1 tsp light soy sauce
2 tsp Shaoxing wine
1/2 tsp sugar
1 tsp cornflour

Method

1. Soak the dried tangerine peel in water until soft. Scrape off the pith. Finely shred it.
2. Rinse the pork cheek. Slice at an angle thinly. Add marinade and stir well. Leave it for 1 hour.
3. Rinse and slice the tofu. Arrange on a plate. Put the marinated pork cheek on top. Sprinkle shredded dried tangerine peel and red chillies on top. Steam over high heat for 10 minutes. Serve hot.

1

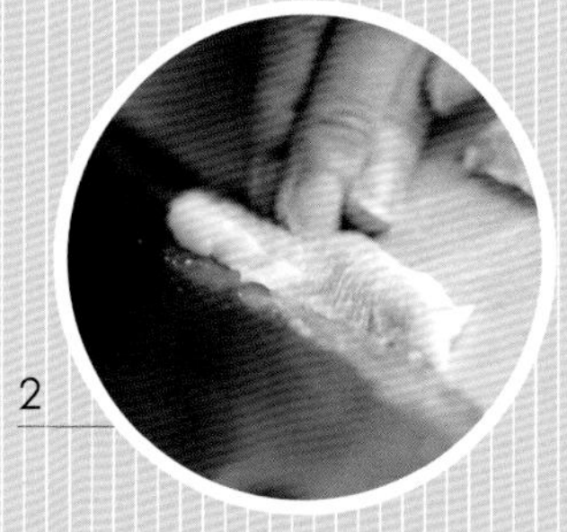
2

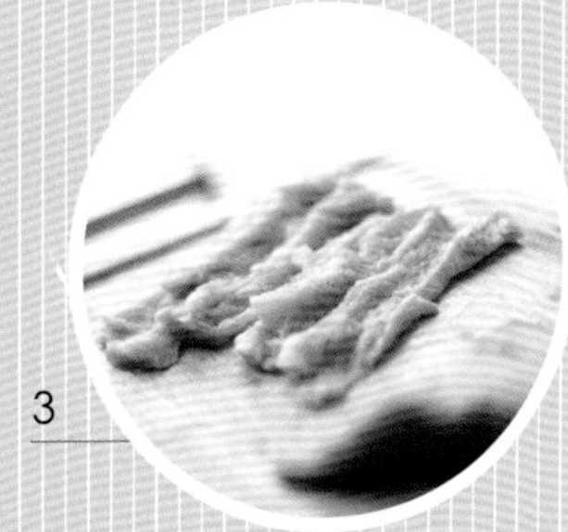
3

零失敗技巧
Successful cooking skills

為甚麼豬頸肉斜切薄片？

由於豬頸肉厚薄不均，難以控制火候，而且直切令肉質粗韌。

Why do you slice the pork cheek at an angle?

The pork cheek has uneven thickness, which makes it unevenly cooked used straight away. Slicing it with a knife vertically cannot cut across its grain and it would end up tough and rubbery.

可用普通軟豆腐嗎？

普通軟豆腐質地軟綿，蒸後容易散開來；布包豆腐則沒有這個問題。

Can I use regular soft tofu instead?

Regular soft tofu is too soft and it tends to fall apart after being steamed. Cloth-wrapped tofu is equally smooth in texture but it stays in one piece after being cooked.

購買新鮮豬頸肉？還是急凍的？

新鮮豬頸肉食味佳，爽口嫩滑，但價錢較昂貴，而且數量有限。

Should I get fresh pork cheek? Or should I get frozen ones?

Fresh ones taste better with its crunchiness and tenderness. But they are more pricey and they are in limited supply.

黑椒百合牛柳條

Stir-fried Beef Tenderloin and Lily Bulb in Black Pepper Sauce

材料
急凍牛柳 250 克
鮮百合 1 個（大）
芥蘭莖 100 克（切度）
紅蘿蔔半個
蒜茸、乾葱茸各 2 茶匙
黑椒碎 1/4 茶匙

醃料
糖半茶匙
生抽、紹酒及粟粉各 1 茶匙
水及油各 1 湯匙
胡椒粉及麻油各少許

調味料
鹽半茶匙

做法
1. 急凍牛柳放於雪櫃下層自然解凍，洗淨，抹乾，順橫紋切成厚件，再切成條狀，下醃料拌勻（最後下油）。
2. 芥蘭飛水，過冷河；紅蘿蔔去皮、切片。
3. 鮮百合剝成瓣狀，切去焦黃部分。
4. 燒熱少許油，下蒜茸、乾葱茸及黑椒碎略炒，放入牛柳條略煎片刻，反轉再煎片刻，略炒推散至五成熟，盛起。
5. 下少許油，放入芥蘭、紅蘿蔔及百合快速兜炒，加入水 2 湯匙，加蓋焗煮 2 分鐘。最後快速拌入牛柳條及調味料，炒勻上碟享用。

Ingredients
250 g frozen beef tenderloin
1 large fresh lily bulb
100 g kale stem (cut into sections)
1/2 carrot
2 tsp finely chopped garlic
2 tsp finely chopped shallot
1/4 tsp crushed black pepper

Marinade
1/2 tsp sugar
1 tsp light soy sauce
1 tsp Shaoxing wine
1 tsp cornflour
1 tbsp water
1 tbsp oil
ground white pepper
sesame oil

Seasoning
1/2 tsp salt

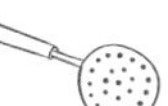
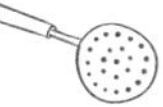

Method

1. Defrost the beef tenderloin in the lower chamber of the refrigerator. Rinse and wipe dry. Cut across the grains into chunks. Then cut into strips. Mix with the marinade (add oil at last).
2. Scald the kale. Rinse with cold water. Peel and slice the carrot.
3. Tear the lily bulb. Cut away the brown edge of the petals.
4. Heat up a little oil. Slightly stir-fry the garlic, shallot and crushed white pepper. Put in the beef tenderloin strips and fry for a moment. Flip over and fry for a while. Slightly stir-fry and scatter until they are rare. Set aside.
5. Put in a little oil. Stir-fry the kale, carrot and lily bulb swiftly. Add 2 tbsp of water. Put a lid on and cook for about 2 minutes. Stir in the kale and seasoning quickly. Stir-fry evenly and serve.

零失敗技巧
Successful cooking skills

如何將急凍牛柳條炒得嫩滑？

有以下幾個要點：一．解凍得宜；二．順橫紋切件，切斷筋鍵；三．快速拌炒！

How to make frozen beef tenderloin strips gentle and soft in stir-fries?

Here are some key points: 1. Defrost properly; 2. Break the tendon by cutting across the grains of the meat into pieces; and 3. Stir-fry quickly!

黑胡椒是此菜之靈魂，如何選購？

選味道濃郁的黑胡椒，使用前才磨成黑椒碎，能保存胡椒香氣！

Black pepper is the soul of this dish. How to select?

Choose those that smell strong. Grind the black pepper right before using to keep the aroma.

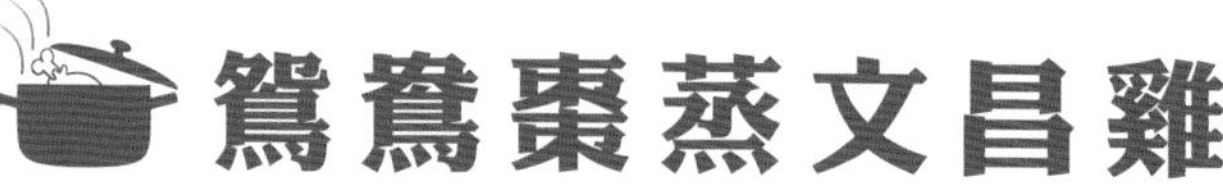

鴛鴦棗蒸文昌雞

Steamed Wenchang Chicken with Black and Red Dates

材料

文昌雞 1 隻（約 1.5 斤至 1 斤 12 兩）
紅棗 8 粒
南棗 4 粒
金針半兩
雲耳 1/4 兩
薑絲半湯匙
葱絲 1 湯匙

調味料

鹽半茶匙
生抽 1 湯匙
老抽 1 茶匙
薑汁酒 1 湯匙
麻油及胡椒粉各少許
生粉 1 湯匙
油半湯匙

做法

1. 文昌雞洗淨、抹乾，斬件。
2. 紅棗及南棗洗淨、去核；金針用水浸軟、打成結；雲耳浸透、去硬蒂。
3. 雞件、紅棗、南棗、金針、雲耳、薑絲及調味料拌勻，待 10 分鐘。
4. 燒滾水，隔水蒸 10 分鐘至雞件全熟，灑入葱絲，加蓋稍焗即成。

Ingredients

1 Wenchang chicken
(about 900 g to 1.05 kg)
8 red dates
4 black dates
19 g dried lily flower
10 g cloud ear fungus
1/2 tbsp shredded ginger
1 tbsp shredded spring onion

Seasoning

1/2 tsp salt
1 tbsp light soy sauce
1 tsp dark soy sauce
1 tbsp ginger juice wine
sesame oil
ground white pepper
1 tbsp caltrop starch
1/2 tbsp oil

Method

1. Rinse the chicken. Wipe dry. Chop into pieces.
2. Rinse and remove the stones from red and black dates. Soak the dried lily flower in water until soft. Knot the dried lily flower. Soak the cloud ear fungus until soft. Remove the hard stalks.
3. Mix the chicken, red dates, black dates, dried lily flower, cloud ear fungus, shredded ginger with the seasoning. Rest for 10 minutes.
4. Bring water to the boil. Steam the chicken over water for 10 minutes until fully cooked. Sprinkle the spring onion on top. Cover with the lid and rest for a moment. Serve.

零失敗技巧
Successful cooking skills

甚麼是文昌雞？

文昌雞是海南省出產的雞，體型細小，肉質嫩滑，皮薄骨脆，雞味濃郁。

What is Wenchang chicken?

It is a type of chicken from Hainan, China. This variety of small chicken with thin skin and crisp bones has a tender texture and a strong chicken flavour.

用紅棗及南棗蒸雞，食味如何？

雞肉帶濃濃的棗香味，伴飯吃一絕！

What is the taste of chicken steamed with red and black dates?

It is absolutely fabulous to be served with rice! The chicken is sweet and aromatic with an intense fragrance of dates.

金針有何益處？

中菜常用的金針，是先蒸後曬的乾品，高蛋白、高鈣，維他命 A 及 B1 豐富，有安神、利尿及鎮靜神經的功能，配合紅棗及南棗一併食用，增加補益功效。

What is the food value of dried lily flower?

The steamed and then sun-dried lily flower is frequently found in Chinese cuisine. Rich in protein, calcium, Vitamins A and B1, it is diuretic and can calm the nerve. Taking with red and black dates can further nourish the body.

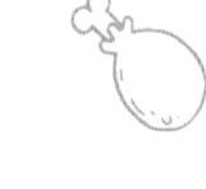

蝦醬炒魷魚筒

Stir-fried Baby Squids in Shrimp Paste

15 分鐘
烹調時間

材料

新鮮小魷魚筒 12 兩
西芹 2 塊
蒜肉 3 粒
蝦醬 1 茶匙
糖半茶匙
紹酒半湯匙

獻汁（拌勻）

粟粉 1 茶匙
水 2 湯匙

做法

1. 西芹撕去老筋，洗淨，切段，飛水備用。
2. 魷魚筒抽出鬚頭及內臟，去掉軟骨，洗淨。
3. 燒熱鑊下油 2 湯匙，下蒜肉及蝦醬炒香，下魷魚筒炒勻，灒酒炒勻，加入西芹及糖炒片刻，最後下獻汁埋獻即成。

Ingredients

450 g fresh baby squids
2 celery stems
3 cloves skinned garlic
1 tsp shrimp paste
1/2 tsp sugar
1/2 tbsp Shaoxing wine

Thickening glaze (mixed well)

1 tsp cornflour
2 tbsp water

Method

1. Tear off the tough veins on the celery. Rinse and cut into short lengths. Blanch in boiling water. Drain.
2. Dress the squids by pulling out their tentacles, head and innards. Remove the bone. Rinse well.
3. Heat a wok and add 2 tbsp of oil. Stir fry skinned garlic and shrimp paste until fragrant. Put in the squids and stir well. Sizzle with wine. Stir further. Add celery and sugar. Mix well. Stir in the thickening glaze and bring to the boil again. Serve.

零失敗技巧
Successful cooking skills

如何撕去西芹老筋？

用手撕出粗絲，將粗纖維拉盡，再用刀削去表皮的纖維。

How do you tear off the tough veins on celery?

Just snap it with your hand and pull off the veins. Then use a peeler to peel off some of the fibres on the skin.

可以原隻魷魚炒嗎？

可以，更可吸收墨汁的營養，做法相同。

Can I stir fry the baby squids in whole?

Yes, you can. You actually get to eat the ink, which is nutritious. In that case, you can skip the dressing step of the squids and cook them in whole. The rest of the method will be the same.

麵豉啫啫唐生菜煲
Sizzling Chinese Lettuce Casserole with Soybean Paste

10 分鐘
烹調時間

材料
唐生菜 12 兩
麵豉醬 1 湯匙
乾葱頭 3 粒（切圈）
薑絲 1 湯匙

做法
1. 唐生菜洗淨，瀝乾水分，備用。
2. 麵豉醬用水 2 湯匙調勻。
3. 瓦煲下油 3 湯匙，用大火燒熱，下乾葱頭及薑絲拌香，加入麵豉醬爆香，煮至油分燒滾，放入唐生菜加蓋焗 3 分鐘，拌勻，趁熱上桌。

Ingredients

450 g Chinese lettuce
1 tbsp fermented soybean paste
3 shallots (cut into rings)
1 tbsp shredded ginger

Method

1. Rinse the Chinese lettuce. Drain and set aside.
2. Mix the fermented soybean paste with 2 tbsp of water.
3. Put 3 tbsp of oil in a casserole. Heat up the oil over high heat. Stir in the shallots and ginger until fragrant. Add the soybean paste and stir-fry until aromatic. When the oil boils, put in the Chinese lettuce. Cover with a lid and cook for 3 minutes. Mix well and serve.

零失敗技巧
Successful cooking skills

甚麼是「啫啫」做法？
烹調時，由於火的熱力令瓦煲的油分燒滾，發出啫啫的聲響而得名。

What is "sizzling" cooking?
When cooking food in a casserole on flame, the heat released makes the oil inside boil and sizzle. It is therefore called sizzling cooking.

必須用瓦煲烹調嗎？
建議用瓦煲烹調，有效保存熱力，提高成功機會。

Need to use a casserole?
It is recommended to use a casserole to retain heat effectively and make it work.

烹調時有何注意之處？
必須留意火候，尤其加入唐生菜後，不要久煮，以免生菜變黃。

What should be noted during cooking?
The heat must be manipulated carefully, especially after putting in the Chinese lettuce. Do not overcook to avoid the Chinese lettuce turning yellow in colour.

椒鹽雞中翼
Fried Mid-joint Chicken Wings with Peppered Salt

材料

急凍雞中翼 8 隻
椒鹽 1 茶匙
紹酒半湯匙

醃料

鹽半茶匙

做法

1. 雞翼解凍，洗淨，抹乾水分，下醃料拌勻醃 1 小時。
2. 平底鑊下油 1 湯匙，待油溫熱，排上雞翼，加蓋，煎至雞翼金黃及全熟，灒酒拌勻，再煎片刻，上碟，最後灑上椒鹽伴吃。

Ingredients

8 frozen mid-joint chicken wings
1 tsp peppered salt
1/2 tbsp Shaoxing wine

Marinade

1/2 tsp salt

Method

1. Defrost the chicken wings. Rinse and wipe dry. Mix with the marinade and rest for 1 hour.
2. Heat up 1 tbsp of oil in a pan. Line in the chicken wings. Cover the pan. Fry until they are golden and cooked through. Sprinkle with the Shaoxing wine. Fry for a while. Sprinkle the peppered salt on top. Serve.

零失敗技巧
Successful cooking skills

如何去掉急凍雞翼之雪味？
以紹酒及椒鹽調味，酒香椒鹽濃，可去除冷藏雪味。

Frozen food carries a characteristic stale taste. How can you remove it?
Season the food with Shaoxing wine and peppered salt. Both condiments have strong smell and aromatic taste which cover up the stale taste.

雞翼灒紹酒後，再煎一會有何好處？
灒紹酒後再煎一會，目的是讓水氣散掉，令雞翼更香脆可口。

Why fry the chicken wings further after sprinkling with Shaoxing wine?
This is to take the moisture out making them crunchier and toothsome.

如何炮製椒鹽？
白鑊炒香幼鹽，再拌入花椒粉即可。

How to make peppered salt?
Stir-fry table salt without oil until fragrant and then mix in ground Sichuan peppercorn.

蕉葉烤魚

Baked Sole Fillet with Banana Leaf

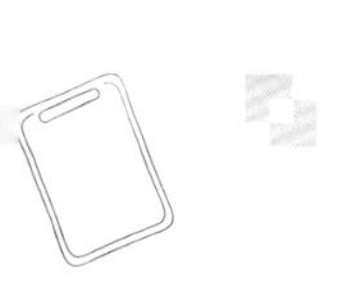

材料

龍脷柳 2 件（8 兩）
乾葱茸 1 湯匙
蒜茸 1 湯匙
蕉葉 1 張
青檸 1 個（切角）

醃料

鹽半茶匙
胡椒粉半茶匙
魚露 2 茶匙
油 2 茶匙

做法

1. 龍脷柳解凍，洗淨，吸乾水分，加入醃料、乾葱茸及蒜茸醃 15 分鐘。
2. 蕉葉裁剪成 30 厘米 x30 厘米正方形（共 2 張），洗淨，飛水，取出備用。
3. 魚柳放於蕉葉（啞色面），摺入蕉葉，包成長方形，用牙籤固定兩端，包成兩份。
4. 將蕉葉魚包放入已預熱之焗爐，用 220℃焗約 18 分鐘至熟透，取出，灑入青檸汁伴食。

1

2

3

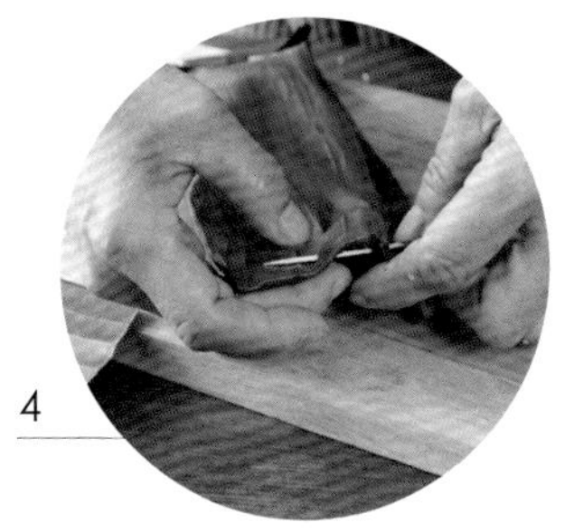
4

Ingredients

2 pieces sole fillet (300 g)
1 tbsp finely chopped shallot
1 tbsp finely chopped garlic
1 banana leaf
1 lime (cut into wedges)

Marinade

1/2 tsp salt
1/2 tsp ground white pepper
2 tsp fish sauce
2 tsp oil

Method

1. Defrost sole fillet. Rinse and wipe dry. Marinate together with finely chopped shallot and finely chopped garlic for 15 minutes.
2. Trim banana leaf into 2 squares of sizes 30 cm by 30 cm. Rinse and scald. Set aside.
3. Put sole fillet on top of the banana leaf (coarse side). Fold up into rectangular shaped and fix the ends with toothpicks. Make 2 sets altogether.
4. Bake the wrapped sole fillet in an oven at 220°C for about 18 minutes until done. Sprinkle over lime juice and serve.

零失敗技巧
Successful cooking skills

哪裏購買蕉葉？若買不到，可用其他葉片代替嗎？
蕉葉於泰式雜貨店有售，價錢相宜，也可用糭葉或荷葉代替。

Where to buy banana leaf? If it cannot be found, can it be replaced with other leaves?
Banana leaf can be bought from Thai grocery stores and its price is reasonable. Or it can be replaced with wrapping leaves for rice dumplings or lotus leaves.

蕉葉必須飛水嗎？
當然，令蕉葉軟身，容易包裹。

Is it a must to scald banana leaf?
Of course. Scalding softens the leaf so that it is easy for wrapping.

焦糖香蔥燒排骨

Caramel Roasted Pork Spareribs

材料

急凍豬肋排 600 克
香茅 1 枝
乾葱茸 3 湯匙
蒜茸 2 茶匙
辣椒乾 1 隻
焦糖魚露汁 1 份（參考 p.51）

做法

1. 急凍豬肋排放於雪櫃下層自然解凍，抹乾備用。
2. 香茅及辣椒乾切碎，加入蒜茸及乾葱茸拌勻，舂爛。
3. 焦糖魚露汁及香茅蒜茸配料拌勻，塗抹肋排醃 2 小時或以上。
4. 將肋排放於烤架（醃汁留用），用上火 200℃燒 30 分鐘至金黃色，掃上醃汁反轉再烤一次，重複以上步驟 2 次，燒至金黃香脆，趁熱享用。

Ingredients

600 g frozen pork spareribs
1 stalk lemongrass
3 tbsp finely chopped shallot
2 tsp finely chopped garlic
1 dried red chilli
1 portion caramel fish sauce (refer to p.51)

Method

1. Defrost the pork spareribs in the lower chamber of the refrigerator. Wipe dry.
2. Chop up the lemongrass and dried red chilli. Mix in the garlic and shallot. Pound with a pestle.
3. Mix the caramel fish sauce with the lemongrass garlic mixture. Spread onto the spareribs and marinate for 2 hours or more.
4. Place the spareribs on a rack (reserve the marinade sauce). Roast at 200℃ on upper heat for 30 minutes until golden. Brush with the marinade sauce. Flip over and roast again. Repeat the steps twice. When it is golden and crispy, serve warm.

焦糖魚露汁

材料

糖 3 湯匙
魚露 2.5 湯匙
乾葱頭 3 粒（切片）
黑椒碎 1/4 茶匙

做法

糖放於鑊內平均分佈，用小火慢慢燒至呈焦黃色，關火，加入餘下的材料煮勻，盛起備用。

Ingredients

3 tbsp sugar
2.5 tbsp fish sauce
3 shallots (sliced)
1/4 tsp crushed black pepper

Method

Put the sugar into a wok evenly. Slowly heat up the sugar over low heat until it turns caramel. Turn off heat. Add the rest ingredients and mix well. Set aside.

零失敗技巧
Successful cooking skills

此道餸有何成功秘訣？
必須掌握煮焦糖的時間，以免焦糖焦燶，令醬汁夾雜苦味，破壞整道菜的效果。

What are the secret tricks behind this recipe?
You must control the cooking time of the caramel. Overcooking the sugar will burn it and make the sauce bitter.

肋排容易入味嗎？
將香茅、辣椒乾、蒜茸及乾葱茸等舂碎後醃製肋排，香味直滲肉內。

Is it easy to make the pork spareribs flavourful?
Pound the lemongrass, dried red chilli, garlic and shallot to marinate the spareribs. Their aroma will penetrate the meat directly.

煎三文魚柳
Fried Salmon Fillets

10分鐘
烹調時間

材料
三文魚柳 2 塊（約 200 克）
檸檬 1 個（切角）

調味料
黑胡椒碎 1 茶匙
幼海鹽 1/3 茶匙

做法
1. 三文魚柳用廚房紙吸乾水分，備用。
2. 平底鑊燒熱橄欖油 2 湯匙，放入三文魚柳用中小火慢煎，灑入調味料，每面各煎約 2 至 3 分鐘，待魚柳呈微黃色上碟，伴檸檬汁享用。

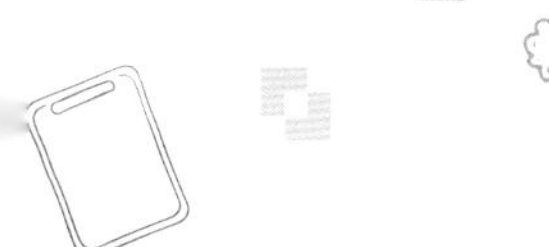

Ingredients

2 salmon fillets (about 200 g)
1 lemon (cut into wedges)

Seasoning

1 tsp chopped black pepper
1/3 tsp fine sea salt

Method

1. Wipe dry the moisture of the salmon fillets with kitchen paper. Set aside.
2. Heat up 2 tbsp of olive oil in a pan. Fry the fillets gently over low-medium heat. Sprinkle with the seasoning. Fry each side for about 2 to 3 minutes. When it turns light brown, dish up. Serve with lemon juice.

零失敗技巧
Successful cooking skills

哪裏購買三文魚柳，質素有保證呢？

日式超市、大型超市或大型急凍食品店，可購買具質素的三文魚柳。

Where to buy quality salmon fillets?

They can be bought at Japanese supermarkets, large supermarkets or frozen food shops.

如何保留三文魚柳之肉汁？

三文魚柳宜用慢火煎香，緊記別煎得過熟，待魚柳中間的肉質呈粉紅色，可保留肉質鮮嫩、軟滑、可口。

How do you keep the salmon moist and succulent?

Fry the salmon over low heat and fry it till it browns nicely on the surface. Do not overcook it. When perfectly cooked, the centre of the fish should still be pink. That's how you can keep it moist, tender and delicious.

家裏沒有橄欖油，可以粟米油代替嗎？

含橄欖香氣的橄欖油，與三文魚最匹配；若家裏甚少使用橄欖油，建議購買小瓶裝橄欖油。

There is no olive oil at home. Can we use corn oil?

Olive oil has the nice aroma of olive and best matches salmon dishes. If it is seldom used, buy a small bottle.

可用粗鹽作調味嗎？

不想購買貴價的價質海鹽，其實日常使用的粗鹽也是海鹽，礦物質含量豐富，略磨後可灑上魚柳使用。

Can we season the salmon with coarse salt?

If you don't want to buy expensive quality sea salt, regular coarse salt will do as it is also sea salt rich in minerals. Roughly crush the salt and sprinkle on the fillets.

豆漿芝士焗西蘭花

Baked Broccoli with Soybean Milk and Cheese

15分鐘
烹調時間

材料

西蘭花半斤
芝士絲 2 湯匙
淡豆漿 100 毫升
粟粉 1 茶匙
蛋白 2 個
鹽 1/4 茶匙

做法

1. 預熱焗爐至 180℃。
2. 西蘭花切成小朵，洗淨，放入淡鹽水浸 30 分鐘，沖洗。
3. 西蘭花放入滾水灼熟，盛起，隔去水分，排於焗盤備用。
4. 淡豆漿、粟粉、蛋白及鹽拌勻，傾入煲內用慢火煮成稀糊狀，澆在西蘭花，灑入芝士絲，放入焗爐焗至芝士表面呈金黃色即成。

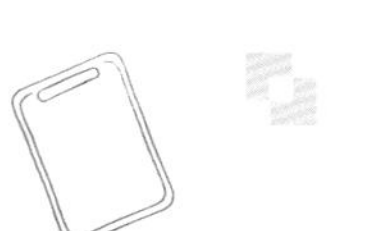
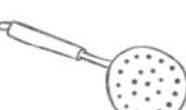

Ingredients

300 g broccoli
2 tbsp shredded cheese
100 ml light soybean milk
1 tsp cornflour
2 egg whites
1/4 tsp salt

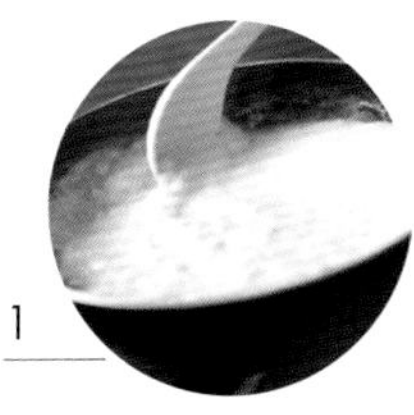
1

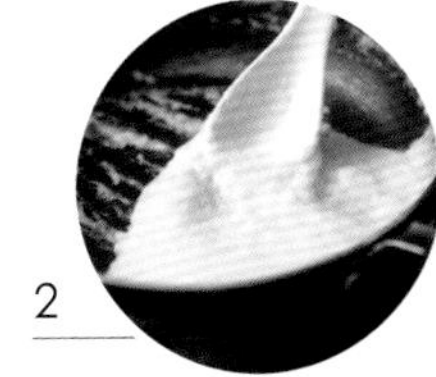
2

Method

1. Preheat an oven to 180°C.
2. Cut the broccoli into small florets. Rinse and soak in light salted water for 30 minutes. Rinse again.
3. Blanch the broccoli until done. Drain and arrange on a baking tray. Set aside.
4. Mix the light soybean milk, cornflour, egg whites and salt well. Cook in a saucepan over low heat until slightly thicken. Pour on the broccoli. Sprinkle the cheese on top. Put in the oven and bake until the surface of the cheese turn golden brown. Serve.

零失敗技巧
Successful cooking skills

如何品嘗嫩滑的炒豆漿？
豆漿料拌勻後，建議用隔篩過濾一次，令豆漿料細緻滑嫩。

How to make the fried soybean milk taste silky-smooth?
It can be done by filtering the soybean milk mixture once with a strainer.

甜豆漿能做出相同的效果嗎？
甜豆漿可做出相同之效果，但須密切留意火候，必須使用內圈火（極小火）慢煮；或可暫離火烹調一會，以防黏底。

Will sweet soybean milk have the same effect?
Yes, it will. But you must closely attend to the heat. Slowly cook on heat from inner ring (extremely low heat), or remove from heat to let it cook for a moment to avoid sticking to the pan.

可用其他蔬菜代替西蘭花嗎？
椰菜花、馬鈴薯、椰菜等均是不錯的選擇。

Can we use other vegetables instead?
Vegetables like cauliflower, potatoes and cabbage are good alternatives.

照燒汁鯖魚
Fried Mackerel in Teriyaki Sauce

材料

急凍鯖魚 2 件
日式照燒汁 4 湯匙
薑絲 1 湯匙
生粉 2 茶匙

醃料

鹽 1/4 茶匙
黑椒粉少許

做法

1. 急凍鯖魚放於雪櫃下層自然解凍，洗淨，抹乾水分，切段，拌醃料待一會。
2. 鯖魚輕輕撲上生粉，備用。
3. 燒熱油 1 湯匙，放入鯖魚煎至兩面金黃色，見鯖魚煎至熟時，加入薑絲同煎。
4. 下日式照燒汁，再略煎至醬汁半收乾，上碟品嘗。

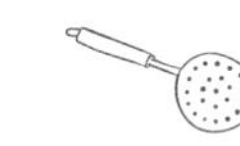

Ingredients

2 pieces frozen mackerel
4 tbsp Japanese teriyaki sauce
1 tbsp shredded ginger
2 tsp caltrop starch

Marinade

1/4 tsp salt
ground black pepper

Method

1. Defrost the mackerel in the lower chamber of the refrigerator. Rinse and wipe dry. Cut into sections. Mix with the marinade and leave for a while.
2. Coat the mackerel thinly with the caltrop starch. Set aside.
3. Heat up 1 tbsp of oil. Fry the mackerel until both sides are golden. When it is almost done, add the ginger and fry together.
4. Pour in the teriyaki sauce. Fry slightly until the sauce is half dry. Serve.

零失敗技巧
Successful cooking skills

煎鯖魚用甚麼火候？

油略燒熱後，放入鯖魚用中火煎透即可。

What heat should be used to fry mackerel?

When oil is heated up, put in the mackerel and fry thoroughly over medium heat.

為何薑絲稍後才放入？

薑絲略煎，足已散發清香薑氣；否則薑絲焦黑，甚至帶焦燶味，會破壞整道菜的香味！

Why put in the shredded ginger later?

Frying the ginger slightly is enough to spread its light aroma; otherwise, it will turn dark in colour and even carry a smell of burnt, spoiling the flavour of the dish.

加入日式照燒汁後，容易焦燶嗎？

應該不會！加入照燒汁後，略煎即可，別煎太久！

Is it easy to get burnt after putting in the teriyaki sauce?

No! After adding the sauce, give a slight frying. Do not fry for too long!

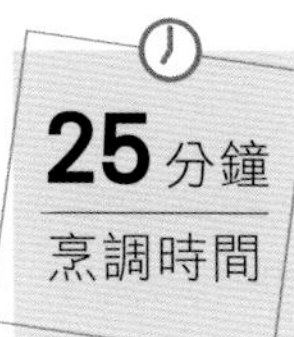

九層塔煮雞球

Three-cup Chicken with Thai Basil

25 分鐘
烹調時間

材料

鮮雞半隻
青、紅甜椒各半個
九層塔 2 棵
薑 10 片
蒜肉 8 粒

三杯醬

麻油、紹酒、生抽各 3 湯匙
冰糖 1 湯匙（舂碎）

做法

1. 青、紅甜椒去籽，洗淨，切塊；九層塔摘取葉片，洗淨備用。
2. 鮮雞洗淨，抹乾水分，斬件。
3. 燒熱鑊下油 2 湯匙，放入甜椒炒片刻，盛起。原鑊下麻油、薑片及蒜肉炒香，加入雞塊、紹酒、生抽及冰糖拌勻煮熟，最後下甜椒及九層塔炒勻即成。

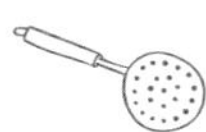

Ingredients

1/2 freshly slaughtered chicken
1/2 red bell pepper
1/2 green bell pepper
2 sprigs Thai basil
10 slices ginger
8 cloves skinned garlic

Three-cup sauce

3 tbsp sesame oil
3 tbsp Shaoxing wine
3 tbsp light soy sauce
1 tbsp rock sugar (crushed)

Method

1. Seed the bell peppers. Rinse well and cut into pieces. Pick the leaves off the basil. Discard the stems. Rinse the basil leaves and set aside.
2. Rinse the chicken and wipe dry. Chop into pieces.
3. Heat a wok and add 2 tbsp of oil. Stir fry the bell pepper briefly. Set aside. In the same wok, add sesame oil, ginger slices and garlic. Stir fry until fragrant. Put in the chicken pieces, Shaoxing wine, light soy sauce and rock sugar. Stir well and cook until chicken is done. Add bell pepper and basil leaves at last. Stir to mix well. Serve.

零失敗技巧
Successful cooking skills

為何使用大量薑片及蒜肉？

薑片及蒜肉等料頭能增添菜式的香味，煮後的薑及蒜更是此菜主角，比雞肉更美味。

Why do you use so much garlic and ginger?

Aromatics like garlic and ginger add fragrance to the dish. In this recipe, the ginger and garlic actually pick up much flavour from the chicken and taste better than the chicken.

只取用九層塔的葉片嗎？

摘取後的九層塔葉片，更能散發獨有香氣。

Why do you use the leaves of the Thai basil only?

The essential oil and volatile fragrance of the basil are released when the leaves are plucked off the stems.

為何用麻油代替生油炒煮雞塊？

三杯醬採用麻油炮製，故一併以麻油炒雞塊，令雞塊散發陣陣麻油香味，令餸菜更香、更惹味，甚至有補身的功效！

Why do you fry the chicken in sesame oil instead of cooking oil?

Sesame oil is the soul of the Three-cup sauce. That's why I fry the chicken in sesame oil to add fundamental warmth in palate that echoes with the sauce. The dish will taste better and smell divine, while having tonifying quality.

紅酒牛柳芥蘭鍋

Beef Tenderloin Casserole with Kale and Red Wine

材料

新鮮牛柳 8 兩
嫩芥蘭半斤
乾葱頭 4 粒（切片）
紅酒半杯
沙茶醬 1 湯匙
紅辣椒 1 隻（切圈）

調味料

鹽 1/4 茶匙
糖半茶匙
清水 1.5 杯

做法

1. 新鮮牛柳洗淨，切薄片。
2. 芥蘭洗淨，飛水，備用。
3. 燒熱砂鍋下油 2 湯匙，下乾葱頭爆香，加入沙茶醬拌炒至香味散發，下調味料煮滾，放入芥蘭，傾入紅酒煮滾，最後下牛柳灼滾及轉色，下紅椒圈略拌，原鍋上桌享用。

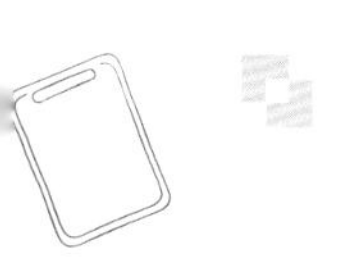

Ingredients

300 g fresh beef tenderloin
300 g young kale
4 shallots (sliced)
1/2 cup red wine
1 tbsp Sa Cha sauce
1 red chilli (cut into rings)

Seasoning

1/4 tsp salt
1/2 tsp sugar
1.5 cups water

Method

1. Rinse and finely slice the tenderloin.
2. Rinse and blanch the kale. Set aside.
3. Heat up a casserole. Put in 2 tbsp of oil. Stir-fry the shallots until scented. Mix in the Sa Cha sauce and stir-fry until the fragrance spreads. Pour in the seasoning. Bring to the boil. Add the kale. Pour in the red wine. Bring to the boil. Blanch the tenderloin until it changes colour. Roughly stir in the red chilli. Serve with the casserole.

零失敗技巧
Successful cooking skills

牛柳配上沙茶醬及紅酒煮成鍋，味道如何？

沙茶醬的味道與紅酒絕對搭配；牛柳用紅酒烹調，令牛肉味香、肉質更嫩。

What is the taste of tenderloin cooked with Sa Cha sauce and red wine?

Sa Cha sauce and red wine match pretty well. The tenderloin cooked in red wine is more fragrant and delicate.

牛柳片需要切得很薄嗎？

盡量將牛柳切成薄片，最後略灼即可，以免煮得太久，肉質太韌。

Need to cut the tenderloin into very thin slices?

Make it as thin as possible for quick blanching. Overcooking makes it tough.

嫩芥蘭容易購買嗎？

市場上有不少嫩芥蘭供應；若購買不到，可以普通芥蘭代替，削去粗硬外皮即可。

Is it easy to buy young kale?

It is often found in the market. If not available, use regular kale and peel the tough skin off.

簡易沙薑汁浸雞髀

Chicken Legs in Spice Ginger Sauce

材料

鮮雞髀 2 隻（約 12 兩）

八角 2 粒

薑 2 片

沙薑粒 4 粒

芫茜 1 棵

調味料（1）

鹽 1 3/4 湯匙

生抽及老抽各半湯匙

糖 3/4 茶匙

清水 3 杯（750 毫升）

調味料（2）

浸雞髀汁 4 湯匙

麻油 1 茶匙

糖 1/3 茶匙

做法

1. 雞髀洗淨，抹乾水分，用竹籤在雞髀厚肉部分大力戳上小孔。
2. 熱鍋下油，下薑片、八角及沙薑粒爆香，傾入調味料（1）煮滾，放入雞髀用慢火煮 15 分鐘，關火，再焗 10 分鐘，浸雞汁留用。
3. 雞髀待涼，斬件，排在碟上。
4. 煮滾調味料（2），澆在雞髀上，以芫茜裝飾即可。

Ingredients

2 fresh chicken legs (about 450 g)

2 star anise

2 slices ginger

4 whole spice ginger

1 stalk coriander

Seasoning (1)

1 3/4 tbsp salt

1/2 tbsp light soy sauce

1/2 tbsp dark soy sauce

3/4 tsp sugar

3 cups water (750 ml)

Seasoning (2)

4 tbsp sauce from soaking chicken legs

1 tsp sesame oil

1/3 tsp sugar

Method

1. Rinse the chicken legs. Wipe dry. Pierce holes in the thick part of the meat with a bamboo skewer heavily.
2. Add oil in a heated pot. Stir-fry the ginger, star anise and spice ginger until aromatic. Pour in the seasoning (1) and bring to the boil. Put in the chicken legs and cook over low heat for 15 minutes. Turn off heat. Rest for 10 minutes with a lid on. Reserve the sauce for later use.
3. When the chicken legs cool, chop into pieces and arrange on a plate.
4. Bring the seasoning (2) to the boil. Sprinkle over the chicken legs. Decorate with the coriander. Serve.

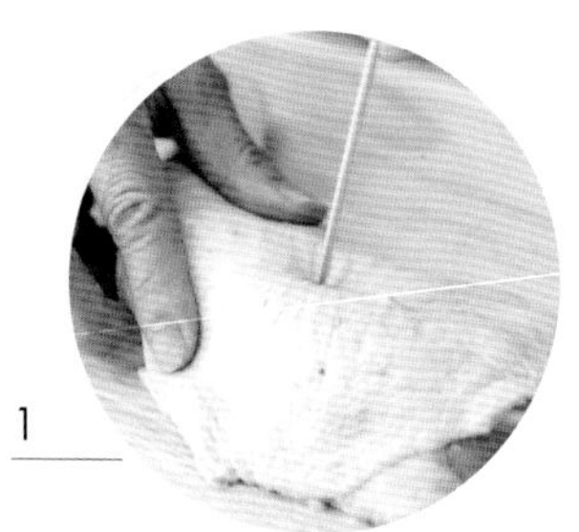

1

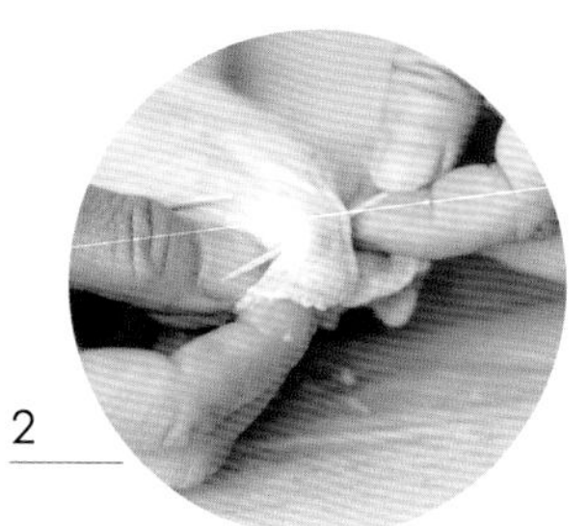

2

零失敗技巧
Successful cooking skills

為何在雞髀上戳上小孔？（圖 1）

令調味香料容易浸入雞肉，美味無窮！

Why pierce holes in the chicken leg? (picture 1)

This is to allow the seasoning penetrate the meat. Yummy!

有何小秘技令雞髀的賣相更美觀？

雞皮受熱後容易收縮捲起，建議用牙籤將雞髀頂部的雞皮及雞肉串起（圖 2），以免雞皮捲縮，令賣相不佳。

Any secret to make the chicken legs look gorgeous?

The chicken skin when heated will contract and curl. It is better to skewer the skin and meat at the top of the chicken leg with a toothpick (picture 2). It will have a better presentation with the skin flattened.

簡單的浸雞髀方法，烹調時間多久？

由於雞髀毋須醃製，而浸煮的過程只需 25 分鐘，輕輕鬆鬆就可品嘗美食，適合繁忙的下班一族。

How long will it take to make this simple dish?

It will take only 25 minutes to make the dish as the marinating process is skipped – simple, easy, and tasty!

枝竹酸菜魚腩煲

Grass Carp with Tofu Stick and Pickled Vegetable in Clay Pot

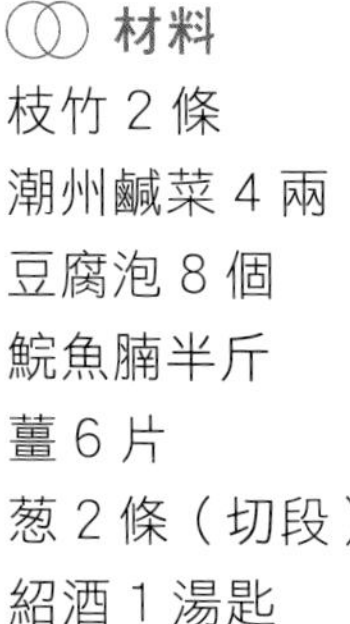

材料

枝竹 2 條
潮州鹹菜 4 兩
豆腐泡 8 個
鯇魚腩半斤
薑 6 片
葱 2 條（切段）
紹酒 1 湯匙

醃料

胡椒粉少許
鹽半茶匙
粟粉 1 茶匙

調味料

麵豉醬 1 湯匙
糖 1 茶匙
水 2 杯

做法

1. 枝竹折段，用水浸軟，隔去水分。
2. 鹹菜切絲，洗淨；豆腐泡飛水，過冷河，擠乾水分。
3. 鯇魚腩洗淨，切成長條形，下醃料拌勻，放入油鑊煎至微黃色，備用。
4. 燒熱瓦鍋，下油 1 湯匙，炒香薑片及葱段，放入鯇魚腩，灒酒，加入鹹菜、枝竹、豆腐泡及調味料煮滾，加蓋，用中火煮 12 分鐘即可。

Ingredients

2 tofu sticks
150 g Chaozhou pickled vegetable
8 deep-fried tofu puffs
300 g grass carp belly
6 slices ginger
2 sprigs spring onion (sectioned)
1 tbsp Shaoxing wine

Marinade

ground white pepper
1/2 tsp salt
1 tsp cornflour

Seasoning

1 tbsp fermented bean paste
1 tsp sugar
2 cups water

Method

1. Break the tofu sticks into sections. Soak in water until soft. Drain.
2. Shred and rinse the pickled vegetable. Scald the deep-fried tofu puffs. Cool in cold water. Squeeze dry.
3. Rinse the grass carp belly. Cut into strips. Mix well with the marinade. Fry in a wok until light brown. Set aside.
4. Heat a clay pot. Pour in 1 tbsp of oil. Stir-fry the ginger and spring onion until fragrant. Add the grass carp belly. Sprinkle with the wine. Put in the pickled vegetable, tofu sticks, deep-fried tofu puffs and the seasoning. Bring to the boil. Cover with a lid and cook over medium heat for 12 minutes. Serve.

零失敗技巧
Successful cooking skills

半斤鯇魚腩可切多少塊？

視乎魚腩之厚度而定，約可切成 3 至 4 塊。

How many pieces can be cut from 300 g of grass carp belly?

It can be cut into 3 to 4 pieces depending on its thickness.

豆腐泡必須飛水嗎？

豆腐泡飛水及過冷河後，可去掉表面的油分，吃起來沒有油膩的感覺。

Must we scald deep-fried tofu puffs?

It is to remove oil on the surface by scalding and rinsing in cold water. They will not taste greasy.

潮州鹹菜的味道很鹹嗎？

不會太鹹，其鹹味適中，只要略沖即可使用。

Is the Chaozhou pickled vegetable very salty?

It is moderately salty. Just slightly rinse before use.

日式咖喱蘋果雞
Japanese Apple Chicken Curry

材料
雞扒 12 兩
日式咖喱（中辣）2 塊
蘋果（小）1 個
馬鈴薯 3 兩
甘筍 3 兩
清水 1.5 杯（375 毫升）

醃料
鹽半茶匙
生抽 2 湯匙
糖 1/4 茶匙
蛋黃 1 個
黑胡椒粉半茶匙
生粉 1 湯匙（後下）

調味料
鹽 1/3 茶匙

做法
1. 雞扒解凍、去皮，洗淨，切件，加入醃料拌勻待 20 分鐘
2. 馬鈴薯及甘筍去皮、切角；蘋果去皮、去心，切角，浸於淡鹽水備用。
3. 熱鑊下油，雞件與生粉拌勻，煎熟盛起。
4. 燒熱油，下馬鈴薯炒至表面略硬，傾入清水及甘筍煮滾，轉慢火煮約 8 分鐘至軟腍，加入咖喱塊煮溶及汁液濃稠，最後下蘋果、雞件及調味料炒勻即成。

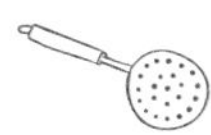

Ingredients

450 g chicken steak
2 pieces Japanese curry (mild)
1 small apple
113 g potato
113 g carrot
1.5 cups water (375 ml)

Marinade

1/2 tsp salt
2 tbsp light soy sauce
1/4 tsp sugar
1 egg yolk
1/2 tsp ground black pepper
1 tbsp caltrop starch (added at last)

Seasoning

1/3 tsp salt

Method

1. Defrost and skin the chicken steak. Rinse and chop into pieces. Mix with the marinade and rest for 20 minutes.
2. Peel the potato and carrot. Cut diagonally. Peel and core the apple. Cut diagonally. Soak in light salted water. Set aside.
3. Add oil in a heated wok. Mix the chicken with the caltrop starch. Fry until done. Set aside.
4. Heat up oil. Stir-fry the potato until the surface hardens. Pour in the water and carrot. Bring to the boil. Turn to low heat and cook for about 8 minutes until tender. Add the curry pieces and cook until they melt and the sauce thickens. Finally put in the apple, chicken and seasoning. Stir-fry evenly. Serve.

零失敗技巧
Successful cooking skills

哪些是日式咖喱？

日式咖喱是盒裝出售，內含 6 小塊，大致可分大辣、中辣及小辣。市面有不同牌子選擇，香辣程度各異，超級市場有售。

Which are the Japanese curries?

Available at supermarkets, Japanese curry is sold in boxes, each containing 6 small pieces. It is majorly divided into hot, mild, and regular flavour. There are many brands of Japanese curry in the market with different degree of spiciness.

馬鈴薯容易煮爛，怎辦？

建議馬鈴薯先用油爆炒至表面略硬，以免最後煮成馬鈴薯茸！

How to keep the potato intact in cooking?

Stir-fry the potato with oil until the surface hardens. This will avoid the potato becoming mash after cooking.

蘋果可預早加入嗎？

以鮮果入饌，通常於最後步驟才加入拌炒，保持蘋果爽脆的口感！

Can we add the apple early?

Fruits are usually added in the final step of cooking to keep their crunchiness.

豆漿鮮淮山雞肉鍋

Yam and Chicken Nabemono with Soybean Milk

25 分鐘
烹調時間

材料

淡豆漿 1.5 杯
雞肉 6 兩
本菇 1 包
鮮淮山 3 兩
薑 4 片
昆布數塊
鰹魚片 2 湯匙
鹽適量

做法

1. 鰹魚片用滾水半杯焗 10 分鐘，隔出湯汁，備用。
2. 本菇切去末端，洗淨；鮮淮山去皮，洗淨，切片；雞肉洗淨、切塊；昆布用濕毛巾抹淨。
3. 淡豆漿及鰹魚湯汁傾入鍋內，加入昆布及薑片煲滾，下雞肉及淮山煮 10 分鐘，加入本菇煮片刻，最後下適量鹽調味。

Ingredients

1.5 cups light soybean milk
225 g chicken meat
1 packet Hon-shimeji mushrooms
113 g fresh Chinese yam
4 slices ginger
a couple piece kelp
2 tbsp dried bonito flakes
salt

Method

1. Soak the dried bonito flakes in 1/2 cup of boiling water covered with a lid for 10 minutes. Filter the soup. Set aside.
2. Cut away the root of the Hon-shimeji mushrooms. Rinse well. Peel, rinse and slice the Chinese yam. Rinse the chicken meat. Cut into pieces. Clean the kelp with a damp towel.
3. Pour the light soybean milk and bonito soup from step 1 into a saucepan. Add the kelp and ginger. Bring to the boil. Add the chicken meat and Chinese yam. Cook for 10 minutes. Put in the mushrooms and cook for a moment. Season with some salt at last and serve.

零失敗技巧
Successful cooking skills

昆布毋須清洗嗎？
只需用濕布略抹，去掉表面鹽分即可。
Do we need to wash kelp?
Just remove the salt on the surface by slightly wiping it with a damp towel.

哪裏購買鰹魚片？
日式超市及大型超市有售，帶陣陣鰹魚香氣。
Where can we buy dried bonito flakes?
They are available at Japanese supermarkets and large supermarkets. They have a nice bonito smell.

淮山的黏液需要洗掉嗎？
淮山的黏液是營養精華所在，切勿洗掉。
Is it necessary to wash away the mucilage of the Chinese yam?
Do not wash it away as it is the nutrition essence.

鮮蝦麻婆豆腐
Stir-fried Tofu with Shrimps

15 分鐘
烹調時間

材料
板豆腐 1 塊
鮮蝦仁 4 兩
免治豬肉 2 兩
薑茸、蒜茸各半湯匙
葱粒 1 湯匙

醃料
胡椒粉少許
生抽半湯匙
粟粉 1 茶匙

獻汁（調勻）
生抽、鎮江醋各半湯匙
麻油、糖各 1 茶匙
粟粉 1 茶匙
水 3 湯匙

做法
1. 鮮蝦仁用滾水灼熟，盛起，隔去水分。
2. 免治豬肉與醃料拌勻；板豆腐切成大粒。
3. 燒熱鑊，下油 1 湯匙，下薑茸及蒜茸炒香，加入免治豬肉拌散，下蝦仁、豆腐粒及水 1/3 杯，煮 5 分鐘，傾入獻汁煮滾，最後拌入葱粒上碟。

Ingredients

1 firm tofu
150 g shelled fresh shrimps
75 g minced pork
1/2 tbsp grated ginger
1/2 tbsp finely chopped garlic
1 tbsp diced spring onion

Marinade

ground white pepper
1/2 tbsp light soy sauce
1 tsp cornflour

Thickening glaze (mixed well)

1/2 tbsp light soy sauce
1/2 tbsp Zhenjiang vinegar
1 tsp sesame oil
1 tsp sugar
1 tsp cornflour
3 tbsp water

Method

1. Blanch the shrimps until done. Drain.
2. Mix the minced pork with the marinade. Coarsely dice the tofu.
3. Heat a wok. Pour in 1 tbsp of oil. Stir-fry the ginger and garlic until fragrant. Add the minced pork and stir evenly. Put in the shrimps, tofu and 1/3 cup of water. Cook for 5 minutes. Pour in the thickening glaze and bring to the boil. Sprinkle with the spring onion at last. Serve.

零失敗技巧
Successful cooking skills

哪類豆腐適合炒煮？
板豆腐的質地較硬實，適合炒煮烹調，以免容易煮爛糊成一團。

Which kind of tofu is suitable for stir-frying?
With a firm texture, the fresh tofu sold at wet markets is a good choice as it is not easily crushed into a mass during the cooking process.

為何先灼熟蝦仁才炒煮？
能夠控制炒煮時間，而且可去掉生蝦之腥味。

Why scald shelled shrimps before stir-frying?
It is to shorten the time of cooking and to get rid of the fishy smell of the shrimps.

三葱焗雞
Stewed Chicken with Assorted Onions

20 分鐘
烹調時間

材料
光雞 8 兩
洋葱半個
乾葱頭 3 粒（開邊）
青葱 2 兩
薑 2 片

醃料
鹽 1/4 茶匙
生抽半湯匙
糖 1/8 茶匙
老抽半茶匙
生粉及油各半湯匙
胡椒粉少許

調味料
水 4 湯匙
蠔油 3/4 湯匙
糖 1/3 茶匙

做法
1. 光雞洗淨，斬件，抹乾水分，加入醃料拌勻待 15 分鐘。
2. 洋葱切粗條；青葱切段，分成葱白及葱段。
3. 熱鑊下油，炒香洋葱備用。
4. 熱鑊下油，下乾葱頭、葱白及薑片爆香，放入雞件炒勻，灒酒，下調味料焗約 8 分鐘，加入洋葱再焗 2 分鐘，最後下青葱段煮半分鐘即成。

Ingredients

300 g chicken
1/2 onion
3 shallots (halved)
75 g spring onion
2 slices ginger

Marinade

1/4 tsp salt
1/2 tbsp light soy sauce
1/8 tsp sugar
1/2 tsp dark soy sauce
1/2 tbsp caltrop starch
1/2 tbsp oil
ground white pepper

Seasoning

4 tbsp water
3/4 tbsp oyster sauce
1/3 tsp sugar

Method

1. Rinse the chicken. Chop into pieces. Wipe dry. Mix with the marinade and rest for 15 minutes.
2. Coarsely cut the onion into strips. Section the spring onion. Separate the white parts from the green.
3. Add oil in a heated wok. Stir-fry the onion until fragrant. Set aside.
4. Add oil in the heated wok. Stir-fry the shallot, white spring onion and ginger until aromatic. Put in the chicken and stir-fry. Sprinkle with Shaoxing wine. Pour in the seasoning and stew for about 8 minutes. Add the onion and stew for 2 minutes. Finally put in the green spring onion. Cook for 1/2 minute and serve.

零失敗技巧
Successful cooking skills

何謂三葱？
三葱指洋葱、乾葱頭及青葱；或用其他葱類（如京葱）代替，也是不錯之選擇。

What are the assorted onions?

They are onion, shallot and spring onion. Using substitutes like Peking scallion is also a good option.

用三葱拌勻雞件，味道如何？
洋葱甜；乾葱頭濃；青葱香，令雞件帶三種葱油香氣！

What is the flavour of chicken mixed with the assorted onions?

The chicken has a mixed fragrance – sweet taste of onion, strong flavour of shallot, and nice smell of spring onion!

如愛吃生洋葱，可省卻炒香洋葱的步驟嗎？
洋葱預先炒香，目的是散發其獨有之甜香味道，建議保留此步驟。

I love to take onion raw. Can I skip the step of stir-frying it?

The purpose of stir-frying the onion is to let its unique sweet flavour spread. Better to keep it.

杞子枸杞菜魚尾湯

Fish Tail Soup with Qi Zi and Wolfberry Vine

30 分鐘
烹調時間

材料

鯇魚尾 12 兩
枸杞菜半斤
杞子 1 湯匙
薑 4 片

做法

1. 枸杞摘好菜葉，洗淨，瀝乾水分。
2. 杞子用熱水洗淨，隔去水分。
3. 鯇魚尾洗淨，抹乾水分，下鹽 1 茶匙抹勻，放入熱油鑊，加入薑片煎至兩面微黃色，隔去油分。
4. 灒入滾水 6 杯，用大火煲半小時，最後加入枸杞菜及杞子煮滾即成。

Ingredients

450 g carp tail
300 g wolfberry vine
1 tbsp Qi Zi
4 slices ginger

Method

1. Tear off and take the leaves from wolfberry vine. Rinse and drain.
2. Rinse Qi Zi with hot water and drain.
3. Rinse carp tail and wipe dry. Rub carp tail evenly with 1 tsp of salt. Fry in a wok with oil and the ginger until both sides turn light brown. Drain.
4. Pour 6 cups of boiling water and boil over high heat for half an hour. Put in wolfberry vine, Qi Zi and bring to boil. Serve.

零失敗技巧
Successful cooking skills

煲魚尾湯有何注意之處？

魚尾湯毋須煲太久，用大火煲半小時味道已足夠；另外枸杞菜及杞子也不宜滾太久，煮滾即可。

Is there anything that needs my attention with this recipe?

Do not boil fish tail soup for too long. Just boil it over high heat for 30 minutes and the flavours should have infused in the soup. Besides, the wolfberry vine and Qi Zi should not be boiled for too long either. Just put them in, bring to the boil again and serve.

用鯇魚尾煲湯，湯水帶草腥味嗎？

先用薑片煎香魚尾，故用鯇魚尾煲出來的湯清甜美味，沒有淡水魚的草腥味。

Does the soup taste fishy?

No, it doesn't. You should fry the fish tail in oil with a slice of ginger which removes fishiness. The soup will end up tasty and light without the grassy fishiness commonly associated with freshwater fish.

節瓜鹹蛋湯

Chinese Marrow Soup with Salted Egg

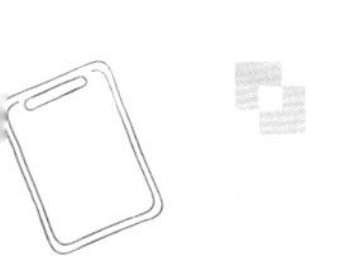
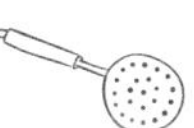
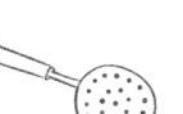

材料

節瓜 2 個
鹹蛋 1 個
瘦肉 4 兩

醃料

生粉半茶匙
鹽少許

做法

1. 節瓜去皮，洗淨，切片。
2. 瘦肉洗淨，切片，用醃料拌勻待片刻。
3. 燒滾清水 8 杯，放入瘦肉和節瓜煲 10 分鐘，加入鹹蛋再煮滾即成。

Ingredients

2 Chinese marrows
1 salted egg
150 g lean pork

Marinade

1/2 tsp caltrop starch
salt

Method

1. Peel the Chinese marrows. Rinse and slice.
2. Rinse and slice the lean pork. Mix with marinade and set aside.
3. Bring 8 cups of water to the boil. Put in the lean pork and Chinese marrows and boil for 10 minutes. Add the salted egg into the soup. Bring to the boil again and serve.

零失敗技巧
Successful cooking skills

如何挑選節瓜？

宜選外表完整、無損傷及體型均稱的節瓜。節瓜肉質爽口，味道清甜，無論炒煮、滾湯也非常適合。

How do you pick the Chinese marrows?

Pick those in whole, with even thickness from end to end, yet without any nicks and dents. Chinese marrows are crisp in texture and lightly sweet in taste. They work fine in stir-fries and quick-boiled soups.

這款湯似乎很容易煮，有何注意？

緊記節瓜毋須煮太久，以免太軟腍；而且加入鹹蛋同煮，所以不必下鹽調味。

This looks like an easy recipe. Is there anything that needs my attention?

Do no cook the Chinese marrows for too long. Otherwise, they'd turn mushy. Besides, you don't need to season the soup with salt as the salted egg yolk should be salty enough.

香芹胡椒豆腐蜆湯

Clam Soup with Tofu and Chinese Celery

15 分鐘
烹調時間

材料

蜆 300 克
芹菜 1 棵
豆腐 1 磚
胡椒粒 1 湯匙
薑 2 片
魚湯或上湯 3.5 杯

調味料

魚露 2 茶匙
鹽適量
胡椒粉適量

做法

1. 蜆用沸水灼至半開口，盛起。
2. 芹菜洗淨，切段；豆腐切小粒。
3. 燒熱少許油，加入胡椒粒略炒，倒入湯、薑片煮滾。
4. 加入蜆、豆腐粒和芹菜，煮至大滾，加入調味料拌勻即可享用。

Ingredients

300 g clams
1 sprig Chinese celery
1 cube tofu
1 tbsp whole peppercorns
2 slices ginger
3.5 cups fish stock or stock

Seasoning

2 tsp fish sauce
salt
ground white pepper

Method

1. Boil clams until the shells half-opened. Remove and set aside.
2. Rinse Chinese celery and cut into sections. Dice tofu
3. Heat oil and stir fry peppercorns briefly. Add stock and ginger and bring to boil.
4. Add clams, tofu and Chinese celery. Bring to boil and add seasoning. Serve.

零失敗技巧
Successful cooking skills

先將蜆灼至半開口，有何好處？
可挑出已死的蜆，而且最後煲煮時，時間毋須太久。
Why do you blanch the clams until the shells are half-open?
You can then fetch and screen those dead ones. This step also helps pre-cook the clams and shortens the cooking time at last.

胡椒粒需要先略炒？
輕輕爆炒胡椒粒，令胡椒味散發出來。
Why do you toast the peppercorns briefly before use?
Gently toasting the peppercorns over low heat helps release their aroma better.

這款湯適合下班後烹調嗎？
絕對合適！做法簡易，而且包含蜆、菜及豆製品，配米飯享用，可品嘗豐富的湯餸晚餐。
Is it a good dinner recipe after a busy day of work?
Absolutely. It's easy and it contains animal protein, vegetables and soy products. It goes well with steamed rice and makes a great soup-meal after work.

紫菜蘿蔔肉碎味噌湯

Miso Soup with Seaweed, White Radish and Pork

材料

白蘿蔔半斤
免治豬肉 4 兩
日式味噌 1.5 湯匙
（用凍開水 3 湯匙調勻）
無砂紫菜 1/4 張（煲湯用）
薑 2 片
葱粒適量
水 6 碗

醃料

生抽 2 茶匙
粟粉 1 茶匙
胡椒粉少許

做法

1. 免治豬肉用醃料拌勻，待片刻。
2. 白蘿蔔去皮，洗淨，切粗條。
3. 燒滾清水 6 碗，加入薑片及蘿蔔煲滾，用中火煲 8 分鐘，加入免治豬肉，傾入味噌煮 7 分鐘，最後下紫菜及葱粒即成。

Ingredients

300 g white radish
150 g minced pork
1.5 tbsp Japanese miso
(mix with 3 tbsp cold drinking water)
1/4 sheet seaweed (for soup)
2 slices ginger
finely chopped spring onion
6 bowls water

Marinade

2 tsp light soy sauce
1 tsp cornflour
ground white pepper

Method

1. Mix minced pork with marinade.
2. Peel white radish, rinse and cut into thick strips.
3. Bring 6 bowls of water to boil. Add ginger and white radish and bring to boil. Boil over medium heat for 8 minutes. Mix in minced pork and miso and boil for 7 minutes. Add seaweed and spring onion. Serve.

零失敗技巧
Successful cooking skills

煲湯之無砂紫菜，哪裏有售？

於普通雜貨店有售，請説明「無砂」紫菜為佳。

Where is sand-free seaweed for making soups available?

It is available in the gorocery store. Choose the sand-free seaweed is the best.

味噌可直接加進爐火內煮嗎？

建議先用凍開水調勻溶解，以免直接加進煲內結成塊狀。又或將味噌放進濾網，在湯內慢慢煮溶。

Can I put the miso in the soup straight away?

It's advisable to thin the miso with cold drinking water first. Miso is too dense in texture and it may stay in one piece without dissolving properly if added straight into the pot. Alternatively, you can put miso in a wire mesh strainer. Put it over the soup and slowly stir the miso through the wire mesh to dissolve it.

木棉魚大豆芽番茄湯

Big-eye Fish Soup with Soybean Sprout and Tomato

30 分鐘
烹調時間

材料

木棉魚 1 條
大豆芽 200 克
番茄 4 個
紅蘿蔔 1 個
薑數片

調味料

鹽適量
胡椒粉少許

做法

1. 木棉魚洗淨，抹乾水分。
2. 大豆芽切去根，洗淨。
3. 番茄去蒂、切件；紅蘿蔔去皮、切件。
4. 燒熱鑊，用薑片擦勻鑊內，下少許油，將木棉魚略煎至微黃色，加入薑片、沸水 5 杯及其餘材料，滾煮約 30 分鐘，加調味料即可飲用。

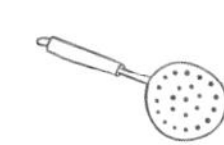
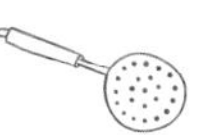

Ingredients

1 big-eye fish
200 g soybean sprouts
4 tomatoes
1 carrot
ginger slices

Seasoning

salt
ground white pepper

Method

1. Rinse big-eye fish and wipe dry.
2. Cut off the roots from soybean sprouts and rinse well.
3. Remove the stalks from tomatoes and cut into pieces. Peel carrot and cut into pieces.
4. Heat wok and rub the wok with a slice of ginger. Add oil and fry big-eye fish until slightly browned. Add ginger, 5 cups of boiling water and remaining ingredients. Boil for 30 minutes. Add the seasoning and serve.

零失敗技巧
Successful cooking skills

大豆芽會太寒涼嗎？
大豆芽偏涼，但煲煮時加入薑片，可減低寒涼之氣。

From Chinese medical point of view, will adding soybean sprouts make this soup too Cold in nature?

Soybean sprouts are Cold in nature. But I also add ginger to the soup to balance it out partly, so that it should not be too Cold.

為何用木棉魚作湯？
木棉魚肉質結實，經滾煮後魚肉不會太爛，可作為餸菜享用。

Why do you use big-eye fish for this soup?

Big-eye fish has firm flesh and is less likely to break down after boiling in soup. You can thus serve its flesh as a dish afterwards.

芫茜皮蛋魚片湯
Grass Carp Soup with Coriander and Century Egg

25 分鐘
烹調時間

材料

鯇魚肉 200 克
皮蛋 1 個
草菇 80 克
免治豬肉 80 克
芫茜 80 克
薑絲 1 湯匙

醃料 1

胡椒粉、麻油各少許

醃料 2

鹽、糖各半茶匙
生抽、粟粉各半茶匙
紹酒半茶匙

調味料

生抽 1 茶匙
鹽適量
胡椒粉少許

做法

1. 鯇魚肉切雙飛，用醃料 1 醃一會。
2. 草菇洗淨、飛水，切半。
3. 免治豬肉加入醃料 2 拌勻。
4. 芫茜洗淨，切度。
5. 皮蛋去泥灰，洗淨，去殼，切成四份。
6. 燒滾水 4 杯，加入免治豬肉、草菇和薑絲，用大火滾後，加入餘下的材料，待再滾，下調味料即可享用。

Ingredients

200 g grass carp meat
1 century egg
80 g straw mushrooms
80 g minced pork
80 g coriander
1 tbsp shredded ginger

Marinade 1

ground white pepper
sesame oil

Marinade 2

1/2 tsp salt
1/2 tsp sugar
1/2 tsp light soy sauce
1/2 tsp cornflour
1/2 tsp Shaoxing wine

Seasoning

1 tsp light soy sauce
salt
ground white pepper

Method

1. Butterfly the fish meat. Mix well with marinade 1.
2. Rinse straw mushrooms, scald and cut into halves.
3. Mix minced pork with marinade 2.
4. Rinse coriander and cut into sections.
5. Clean and rinse century egg. Shell and cut into quarters.
6. Bring 4 cups of water to boil. Add minced pork, straw mushrooms and ginger. Bring to the boil over high heat. Add the remaining ingredients. Bring to the boil again. Sprinkle with the seasoning and serve.

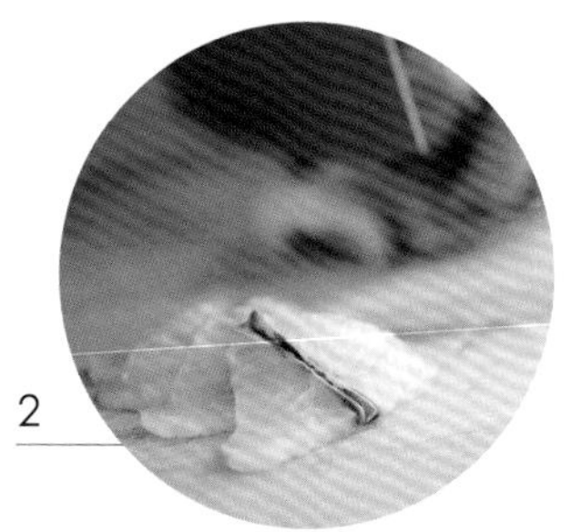

零失敗技巧
Successful cooking skills

魚肉切雙飛有何技巧？

用利刀切入魚肉位置，第一刀不要切斷，第二刀切開即成雙飛狀，熟後的魚肉較美觀。

What are the tricks when cutting the fish into butterflied slices?

When you make the first cut, do not cut all the way through. Then cut all the way in the second cut. The fish should look like two thin slices connected at the centre. It would look bigger and nicer that way.

此湯可配甚麼一同品嘗？

芫茜味清香，可配粉麵或米飯同吃，是下班後完美的湯餸配搭。

What should I serve this soup with?

Coriander is aromatic in smell. You may serve this with rice or noodles. It's the perfect meal after work.

蝦仁雪耳羹

Shrimp and White Fungus Thick Soup

材料

免治豬肉 3 兩
鮮蝦 8 兩
雪耳 1 朵
蛋白 2 個（拌勻）
水 4 碗

醃料

生抽 2 茶匙
胡椒粉少許
粟粉 1 茶匙
水 2 湯匙

調味料

鹽半茶匙
胡椒粉少許

獻汁（調勻）

馬蹄粉 2 湯匙
水 4 湯匙

做法

1. 雪耳用水浸 2 小時，剪去硬蒂，飛水，過冷河，切碎。
2. 鮮蝦去殼，挑腸，洗淨，抹乾水分，備用。
3. 免治豬肉與醃料拌勻。
4. 煮滾水 4 碗及雪耳碎，加入肉碎拌勻煮滾，下蝦仁煮滾片刻，下調味料拌勻，轉小火，加入獻汁（一邊倒入一邊拌至合適的濃度），熄火，最後加入蛋白拌勻即成。

Ingredients

113 g minced pork
300 g fresh shrimps
1 head white fungus
2 whisked egg whites
4 bowls water

Marinade

2 tsp light soy sauce
ground white pepper
1 tsp cornflour
2 tbsp water

Seasoning

1/2 tsp salt
ground white pepper

Thickening glaze (mixed well)

2 tbsp water chestnut powder
4 tbsp water

Method

1. Soak white fungus for 2 hours, cut off the hard stems. Scald, rinse and chop into pieces.
2. Shell and de-vein shrimps, rinse and wipe dry.
3. Mix minced pork with marinade.
4. Bring white fungus in 4 bowls of water to boil. Add minced pork and bring to the boil. Put in shrimps and cook a while. Mix in seasoning. Turn to low heat. Stir in thickening glaze until the desired thickness is achieved. Turn off heat and mix in egg whites. Serve.

零失敗技巧

Successful cooking skills

用蛋白埋獻，如何才能做得綿滑？

必須熄火後才下蛋白拌勻，蛋白軟滑香綿，口感不會太粗。

How to make the egg white smooth in the thick soup?

Before adding egg white, you should turn off the heat first, so it will become smooth and soft, not hard and stiff.

做湯羹有何技巧？

建議所有材料切得細碎及小塊，而且拌入肉碎後弄散，會品嘗細緻的湯羹質感。

Is there any trick behind this recipe?

Chop all ingredients finely. When you stir in the minced pork, it will break into tiny bits. The thick soup would then be velvety smooth.

芥菜排骨湯
Sparerib Soup with Mustard Greens

材料

排骨半斤
芥菜 1 棵
薑 1 片

Ingredients

300 g spareribs
1 stalk mustard greens
1 slice ginger

做法

1. 排骨洗淨，飛水，洗淨。
2. 芥菜洗淨，切塊。
3. 燒滾清水 6 杯，放入排骨和薑煲 30 分鐘，加入芥菜煲至熟透，下鹽調味即成。

Method

1. Rinse and scald the spareribs. Drain and rinse well.
2. Rinse the mustard greens and cut into pieces.
3. Bring 6 cups of water to the boil. Put in the spareribs and ginger and boil for 30 minutes. Put in the mustard greens and boil until done. Season with salt. Serve

零失敗技巧
Successful cooking skills

排骨如何飛水？

凍水時放入排骨，煲滾後煮 10 分鐘，令肉內的血水及污垢容易逼出來，令煲出來的湯水較清澈。

How do you blanch the spareribs?

Put the ribs into a pot of cold water. Then bring to the boil and cook them for 10 minutes. This way, the blood and scum can be drained more thoroughly from the pork. The soup will end up clearer.

如何挑選芥菜？

選大棵的芥菜，作為湯餸有豐富的口感。

How do you pick mustard greens?

Pick those that are bigger in sizes. They make a great dish with crunchy texture after cooked.

韓式海鮮湯
Korean Seafood Soup

25 分鐘
烹調時間

材料

中蝦 8 隻
蜆 300 克
大豆芽 100 克
京葱 1 棵
鹹蝦仔 2 茶匙
薑數片
麻油 1 湯匙
韓式辣醬 1 湯匙

調味料

鹽適量
醬油 2 茶匙

做法

1. 中蝦及蜆洗淨。
2. 京葱斜切；大豆芽沖淨。
3. 燒熱麻油，將中蝦略煎盛起。加入薑、鹹蝦仔和韓式辣醬炒片刻，下大豆芽、水 3.5 杯煮約 10 分鐘，加入蜆，待蜆開口，下其餘材料和調味料即可享用。

Ingredients

8 prawns
300 g clams
100 g soybean sprouts
1 sprig Peking scallion
2 tsp salted tiny shrimps
ginger slices
1 tbsp sesame oil
1 tbsp Korean chilli paste

Seasoning

salt
2 tsp soy sauce

Method

1. Rinse the prawns and clam shells.
2. Cut the scallion diagonally. Rinse the soybean sprouts.
3. Heat sesame oil and briefly fry prawns and remove. Stir fry ginger, salted tiny shrimps and Korean chilli paste. Add soybean sprouts and 3.5 cups of water and cook for 10 minutes. Add clams and cook until the shells are fully opened. Add the remaining ingredients and seasoning. Serve.

零失敗技巧
Successful cooking skills

鹹蝦仔哪裏有售？

鹹蝦仔於雜貨店有售，體積雖小，但鹹香味濃郁。

Where can I buy the salted tiny shrimps?

Salted tiny shrimps are available in grocery stores. Although they are small in size, they give a sufficient amount of tanginess.

如何令海鮮湯更濃郁好味？

建議先將蝦、鹹蝦仔及韓式辣醬等材料炒香，煲煮後鮮味充足，惹味可口。

How do you make this soup richer and tastier?

Stir-fry the shrimps, salted tiny shrimps and other seafood in Korean chilli paste until fragrant first. That would bring out the flavours of the seafood and make the soup tastier.

菠菜豬膶湯
Pork Liver Soup
with Spinach

材料

豬膶 4 兩
菠菜 12 兩
薑 2 片

醃料

薑汁 1.5 茶匙
生粉、生抽各 1 茶匙

做法

1. 豬膶洗淨，切片，加入醃料拌勻，醃片刻。
2. 菠菜洗淨，切段。
3. 燒滾清水 5 杯，放入全部材料，加入少許油煲滾，下鹽調味，再煲片刻即成。

Ingredients

150 g pork liver
450 g spinach
2 slices ginger

Marinade

1.5 tsp ginger juice
1 tsp caltrop starch
1 tsp light soy sauce

Method

1. Rinse and slice the pork liver. Add marinade. Mix well and set aside.
2. Rinse the spinach and cut into sections.
3. Bring 5 cups of water to the boil. Add all ingredients and a small amount of oil and bring to the boil. Season with salt and boil for a while. Serve.

零失敗技巧

Successful cooking skills

煲煮豬膶有何技巧？

豬膶不宜久煮，否則質感粗糙，口感不太好吃。

Is there any skill in cooking pork liver?

Do not boil the pork liver for too long. Otherwise, it would turn rubbery and tough.

怕豬膶的腥味，怎辦？

試試用薑汁醃一下豬膶，可去掉不當的氣味。

I'm not a big fan of pork liver because of the gamey taste. Is there anything I can do to make taste less strong?

You can marinate the pork liver with ginger juice. That would remove the gamey taste.

苦瓜鹹菜海魚湯
Fish Soup with Bitter Melon and Salted Mustard Green

20 分鐘
烹調時間

材料

小海魚 12 兩
苦瓜半斤
鹹菜 4 兩
薑 3 片

做法

1. 苦瓜開邊，去籽，洗淨，切片。
2. 鹹菜洗淨，切絲。
3. 海魚劏好，洗淨，抹乾水分，放入熱油鑊內，下薑片煎至海魚兩面微黃色，隔去多餘油分，灒入滾水 5 杯，用大火煮 10 分鐘，加入苦瓜及鹹菜以大火滾 5 分鐘即成，上碟享用。

Ingredients

450 g small sized seawater fish
300 g bitter melon
150 g salted mustard green
3 slices ginger

Method

1. Cut bitter melon into halves and remove seeds. Rinse and slice.
2. Rinse and shred salted mustard green.
3. Cut open and gut fish. Rinse and wipe dry. Fry in a wok with oil and ginger until both sides turn light brown. Drain excessive oil and pour in 5 cups of boiling water. Boil over high heat for 10 minutes. Put in bitter melon and salted mustard green. Boil over high heat for 5 minutes. Serve.

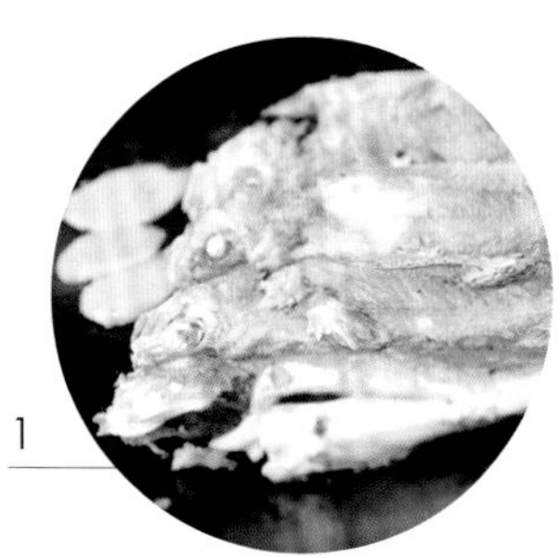

1

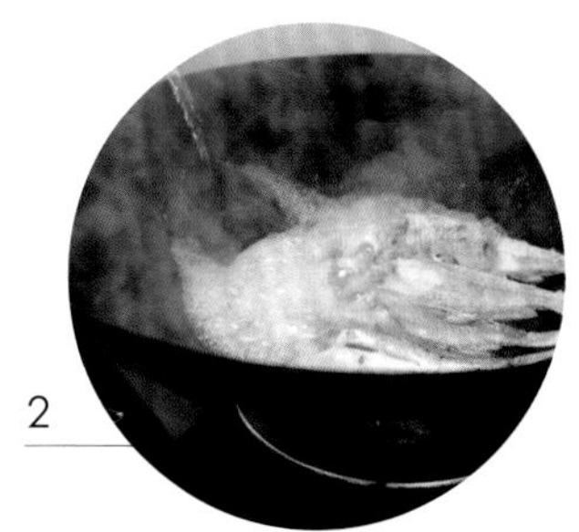

2

3

零失敗技巧
Successful cooking skills

購買哪些小海魚熬湯？

任何海魚皆可，將海魚煎透後灒入滾水，可煲成奶白色的魚湯。

What kind of fish should I use?

Basically, you can use every kind of seawater fish. To make milky fish soup, pour in boiling water after frying the fish.

用苦瓜煲成魚湯，有何功效？

此湯清熱降火、潤腸排毒，適合捱更抵夜的上班一族。

What is the medicinal value of this soup?

From Chinese medical point of view, it clears Heat, subdues Fire, lubricates the intestines and detoxifies. It is a great tonic for those who need to work till late hours at night.

紅蘿蔔竹蔗馬蹄湯

Water Chestnut Soup with Carrot and Sugarcane

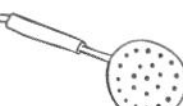

材料

馬蹄 10 粒
瘦肉 4 兩
紅蘿蔔 1 個
竹蔗 1 紮

做法

1. 瘦肉洗淨，飛水，盛起洗淨。
2. 竹蔗洗淨，每條斬開四件。
3. 馬蹄去皮，洗淨。
4. 紅蘿蔔去皮，洗淨，切塊。
5. 燒滾清水 8 杯，放入全部材料煲 1 小時，下鹽調味即成。

Ingredients

10 water chestnuts
150 g lean pork
1 carrot
1 packet sugarcane

Method

1. Rinse and scald the lean pork. Drain and rinse well.
2. Rinse and cut each sugarcane into four sections.
3. Peel and rinse the water chestnuts.
4. Peel and rinse the carrot. Cut into pieces.
5. Bring 8 cups of water to the boil. Put in all ingredients and boil for 1 hour. Season with salt and serve.

零失敗技巧
Successful cooking skills

如何選購馬蹄？

購買完整、未去皮的馬蹄，買回來後才去皮，質素較佳，別購買市面上已去皮之馬蹄。

How to choose the water chestnuts?

It is better to buy water chestnuts with peel them by yourself. Do not buy skinned water chestnuts.

此湯對身體有何好處？

竹蔗及馬蹄清熱潤肺、滋潤身體，適合春夏兩季飲用。

Is this soup good for the body?

It is a great soup for spring and summer. From Chinese medical point of view, sugarcane and water chestnuts clear Heat, moisten the Lungs and moisturize the skin.

洋葱番茄薯仔牛肉湯

Beef Soup with Onion, Tomato and Potato

30 分鐘

烹調時間

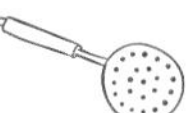

材料

洋葱 1 個
番茄 3 個
馬鈴薯 2 個
免治牛肉 4 兩
鹽半茶匙
油半湯匙

做法

1. 洋葱去外衣，洗淨，切絲；番茄洗淨，去蒂，切角；馬鈴薯去皮，洗淨，切塊。
2. 燒熱鑊下油，加入洋葱炒香，下番茄炒勻，注入熱水 6 碗及薯仔煲滾，再煲 25 分鐘，加入免治牛肉拌散，煲滾後下鹽調味即成。

Ingredients

1 onion
3 tomatoes
2 potatoes
113 g minced beef
1/2 tsp salt
1/2 tbsp oil

Method

1. Peel onion, rinse and shred. Rinse tomatoes, remove stalks and cut into wedges. Peel potatoes, rinse and cut into pieces.
2. Heat wok and fry onion with oil. Stir in tomatoes. Pour in 6 bowls of hot water, add potatoes and bring to boil. Boil for 25 minutes. Mix in minced beef and bring to boil. Season with salt. Serve.

零失敗技巧
Successful cooking skills

看似簡單的滾湯，有何訣竅？

最重要是將洋葱及番茄炒香，煲出來的湯水與別不同，特別香甜惹味。

This soup seems very easy to make, is there any trick?

Onion and tomatoes should be fried until fragrant beforehand. It will enrich the taste and sweetness of the soup.

免治牛肉毋須醃製嗎？

這款湯品嘗的是原汁原味的蔬菜及肉香，用番茄、洋葱等多款材料煲煮，已經惹味香口。

Should the minced beef be marinated first?

No. This soup is to showcase the authentic flavours of beef and veggies. With tomatoes, onion and other ingredients, this soup is more than flavourful enough.

鹹菜桂花魚湯

Chinese Perch Soup with Salted Mustard Greens

材料

桂花魚 1 條
鹹菜 100 克
紅蘿蔔 1/4 個
芹菜 1 棵
胡椒粒 2 茶匙
薑 2 片
粉絲半小紮

調味料

鹽半茶匙
魚露 2 茶匙

做法

1. 桂花魚去內臟，起肉，切件；魚骨斬件。
2. 紅蘿蔔去皮、切絲；芹菜去根部，切段。
3. 鹹菜用少許淡鹽水浸一會，擠乾水分，切幼絲；粉絲略浸一會。
4. 燒熱少許油，加入魚骨、魚肉略煎至微黃色，下胡椒粒、薑片略炒片刻。倒入沸水 3 杯，再煮 8 分鐘。
5. 加入餘下材料，煮滾後下調味料即可享用。

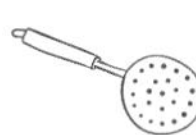

Ingredients

1 Chinese perch
100 g salted mustard greens
1/4 carrot
1 sprig Chinese celery
2 tsp peppercorns
2 slices ginger
1/2 small bundle mung bean vermicelli

Seasoning

1/2 tsp salt
2 tsp fish sauce

Method

1. Gut Chinese perch, separate the meat and cut into pieces. Cut the bone into pieces.
2. Peel and shred carrot; remove the root from Chinese celery and cut into sections.
3. Soak salted mustard greens with lightly salted water, squeeze and cut into thin strips; briefly soak mung bean vermicelli.
4. Heat oil and fry perch bone and meat until slightly browned. Add peppercorns and ginger and stir fry. Pour in 3 cups of boiling water and cook for 8 minutes.
5. Put in other ingredients and bring to boil. Add seasoning. Serve.

零失敗技巧
Successful cooking skills

怎樣令湯水更美味？

魚肉起骨後，保留魚骨作煲魚湯之用，香煎後可煲成奶白色的香甜魚湯。

How do you make this soup even tastier?

After you debone the fish, keep the bones and use it to make soup. Fry the bones in a little oil until golden on both sides. Then pour in boiling water to make the soup milky white and delicious.

鹹菜為何用淡鹽水浸泡？

可去掉鹹菜的鹽分，令湯水鹹味適中。

Why do you soak the salted mustard greens in lightly salted water?

Salted mustard greens could be too salty to be used straight away. But soaking them in freshwater may make them too bland. Thus, soaking them in lightly salted water works best.

冬瓜海鮮湯
Winter Melon and Seafood Soup

材料

冬瓜 12 兩
蝦 2 兩
魚柳 4 兩
乾瑤柱 3 粒
薑 2 片
葱粒適量

做法

1. 冬瓜去皮、去瓤，洗淨，切塊。
2. 蝦、魚柳洗淨，瀝乾水分。
3. 乾瑤柱用水浸軟，撕成細條。
4. 燒滾清水 6 杯，放入薑片和冬瓜煲 15 分鐘，加入海鮮材料，再煲 15 分鐘，最後下鹽和葱粒即成。

Ingredients

450 g winter melon
75 g shrimps
150 g fish fillet
3 dried scallops
2 slices ginger
diced spring onion

Method

1. Peel the winter melon and remove the pith. Rinse and cut into pieces.
2. Rinse the shrimps and fish fillet. Drain.
3. Soak the dried scallops until soft. Tear into shreds.
4. Bring 6 cups of water to the boil. Add the ginger slices and winter melon and boil for 15 minutes. Add the remaining seafood and boil for 15 minutes. Season with salt and sprinkle the spring onion on top. Serve.

零失敗技巧
Successful cooking skills

如何節省烹調成本？

乾瑤柱令湯水帶清甜味道，也可使用小元貝、蝦乾或蝦米煲煮，降低成本。

Is there any way to cut corners with this recipe?

Dried scallops are the key ingredients that impart seafood sweetness to the soup. But if you cook on a budget, use smaller dried scallops or dried baby shrimps to bring the cost down.

可以選用其他海鮮嗎？

當然可以！鮮蜆、帶子、鮮魷等也可加入煲成海鮮湯，悉隨尊便。

Can I use other seafood instead?

Of course you can. Live clams, scallops, or squids work just fine in this soup. Feel free to pick your favourites.

草菇肉絲豆腐羹

Tofu Thick Soup with Straw Mushroom and Pork

材料

板豆腐 1 塊
柳梅肉 4 兩
草菇 2 兩
甘筍絲少許
水 3 碗

醃料

生抽 2 茶匙
粟粉 1 茶匙
胡椒粉少許

調味料

薑汁 1 茶匙
鹽半茶匙

獻汁（調勻）

馬蹄粉 2 湯匙
水 4 湯匙

做法

1. 柳梅肉洗淨，切絲，下醃料拌勻。
2. 草菇削去菇底的污垢，洗淨，飛水，過冷河，切碎。
3. 板豆腐洗淨，壓碎，過濾。
4. 煮滾清水 3 碗，放入柳梅肉拌散煮滾，加入豆腐、草菇及甘筍絲煮 5 分鐘，下調味料煮滾，轉小火，下獻汁（一邊倒入一邊拌勻至合適的濃度），盛起享用。

Ingredients

1 cube firm tofu
150 g pork butt
75 g straw mushrooms
shredded carrot
3 bowls water

Marinade

2 tsp light soy sauce
1 tsp cornflour
ground white pepper

Seasoning

1 tsp ginger juice
1/2 tsp salt

Thickening glaze (mixed well)

2 tbsp water chestnut powder
4 tbsp water

Method

1. Rinse pork butt and shred. Mix well with marinade.
2. Cut off any dirt from straw mushrooms. Rinse, scald, rinse with cold water and chop finely.
3. Rinse tofu, crush into pieces and sift well.
4. Bring 3 bowls of water and mix in pork butt. Bring to boil and add tofu, straw mushrooms and carrot and boil for 5 minutes. Add seasoning and bring to boil. Turn to low heat. Stir in thickening glaze until the desired thickness is achieved. Serve.

零失敗技巧
Successful cooking skills

如何做到滑溜溜的豆腐羹？
秘訣在於豆腐壓碎後過濾，再煮一會，豆腐羹必然嫩滑無比。

How to make a smooth tofu thick soup?
The trick is to sift the tofu after crushing; it will become very smooth after boiling.

淮山紅菜頭葉素菜湯

Vegetable Soup with Beetroot Leave and Yam

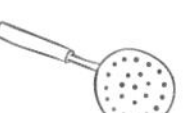

材料

紅菜頭葉 5 至 6 塊
鮮淮山 1 段
洋葱半個
大頭菜 3 片
椰菜、甘筍、西芹各適量
薑 3 片
橄欖油 1 湯匙

做法

1. 鮮淮山洗淨污泥，削去外皮，洗淨，切薄片。
2. 紅菜頭葉洗淨，切碎；大頭菜洗淨，切條。
3. 洋葱去外衣，洗淨，切絲；甘筍去皮，洗淨，切片；椰菜及西芹洗淨，切碎。
4. 凍鑊下橄欖油，加入全部蔬菜（鮮淮山除外）以慢火炒軟，傾入滾水 8 碗用中火煲 10 分鐘，最後下鮮淮山再煲 30 分鐘即成。

Ingredients

5-6 beetroot leaves
1 section fresh yam
1/2 onion
3 slices salted turnip
cabbage
carrot
celery
3 slices ginger
1 tbsp olive oil

Method

1. Rinse off any mud from fresh yam. Peel, rinse and cut into thin slices.
2. Rinse beetroot leaves and chop. Rinse salted turnip and cut into strips.
3. Peel onion, rinse and shred. Peel carrot, rinse and slice. Rinse cabbage and celery and chop.
4. Add olive oil in a cold wok. Add all vegetable (except yam) and stir fry over low heat until soft. Add 8 cups of boiling water and boil over medium heat for 10 minutes. Add fresh yam and boil for 30 minutes. Serve.

零失敗技巧
Successful cooking skills

紅菜頭葉哪裏有售？

要原個紅菜頭買回來，再自行摘取葉片使用，整個紅菜頭的營養也不浪費。

How can I get beetroot leaves?

To prevent wasting beetroots, get whole beetroots from vegetable stores and take the leaves for the soup.

編者
Forms Kitchen編輯委員會

美術設計
馮景蕊

排版
劉葉青

出版者

香港鰂魚涌英皇道1065號
東達中心1305室
電話
傳真
電郵
網址

發行者
香港聯合書刊物流有限公司
香港新界大埔汀麗路36號
中華商務印刷大廈3字樓
電話
傳真
電郵

承印者
中華商務彩色印刷有限公司

出版日期
二零一九年三月第一次印刷

Editor
Editorial Committee, Forms Kitchen

Design
Carol Fung

Typography
Rosemary Liu

Publisher
Forms Kitchen
Room 1305, Eastern Centre, 1065 King's Road,
Quarry Bay, Hong Kong
Tel: 2564 7511
Fax: 2565 5539
Email: info@wanlibk.com
Web Site: http://www.wanlibk.com
http://www.facebook.com/wanlibk

Distributor
SUP Publishing Logistics (HK) Ltd.
3/F., C&C Building, 36 Ting Lai Road,
Tai Po, N.T., Hong Kong
Tel: 2150 2100
Fax: 2407 3062
Email: info@suplogistics.com.hk

Printer
C & C Offset Printing Co., Ltd.

Publishing Date
First print in March 2019

Published in Hong Kong by Forms Kitchen,
a division of Wan Li Book Company Limited.
ISBN 978-962-14-7006-5

鳴謝以下作者提供食譜（排名不分先後）：
黃美鳳、Feliz Chan、Winnie姐